INSIDE
ArcView®

Scott Hutchinson and Larry Daniel

INSIDE ArcView®
By Scott Hutchinson and Larry Daniel

Published by:
OnWord Press
2530 Camino Entrada
Santa Fe, NM 87505-4835 USA

All rights reserved. No part of this book may be reproduced or transmitted in any form or by any means, electronic or mechanical, including photocopying, recording, or by any information storage and retrieval system without written permission from the publisher, except for the inclusion of brief quotations in a review.

The authors gratefully acknowledge permission to reprint two illustrations from *Design with Nature* (Garden City, NY: Doubleday & Co., 1971, pp. 134, 135) by Ian McHarg. Copyright © 1971 by Ian McHarg. Reprinted by permission of Ian McHarg.

The ArcView screen captures are used herein with the permission of Environmental Systems Research Institute, Inc. Copyright © 1994-95 Environmental Systems Research Institute, Inc. All rights reserved.

Copyright © 1995 Scott Hutchinson and Larry Daniel
SAN 694-0269
First Edition, 1995
10 9 8 7 6 5 4 3 2 1
Printed in the United States of America

Library of Congress Cataloging-in-Publication Data

Hutchinson, Scott, 1952-
    Inside ArcView / by Scott Hutchinson and Larry Daniel. -- 1st ed.
      p. cm.
    Includes index.
    ISBN 1-56690-016-6
    Geographic information systems—Computer program. 2. ArcView.
 I. Daniel, Larry, 1961-   II. Title.
G70.2.H87   1995
910' .285'574--dc20   95-9825
                                                                         CIP
                                                                         Rev.

Trademark Acknowledgments

ArcView and ARC/INFO are registered trademarks of the Environmental Systems Research Institute (ESRI), Inc. OnWord Press is a registered trademark of High Mountain Press, Inc. Other terms mentioned in this book that are believed to be trademarks or service marks have been appropriately capitalized. OnWord Press cannot attest to the accuracy of this information. Use of a term in this book should not be regarded as affecting the validity of any trademark or service mark. OnWord Press and the authors make no claim to these marks.

Warning and Disclaimer

This book is designed to provide information on learning how to use the ArcView program. Every effort has been made to make the book as complete, accurate, and up-to-date as possible; however, no warranty or fitness is implied.

The information is provided on an "as-is" basis. The authors and OnWord Press shall have neither liability nor responsibility to any person or entity with respect to any loss or damages in connection with or arising from the information contained in this book.

About the Authors

Since 1983, **Scott Hutchinson** has worked with geographic information systems (GIS) in the U.S. Soil Conservation Service, the Arizona Department of Transportation, and the Arizona State Land Department. He has a B.S. degree in agriculture (soil science) from the University of Arizona. At present, he is using ArcView to develop a system linking digitized land parcel and other resource data to the Arizona State Land Department Business Systems database toward providing a query and mapping system on the state's trust land. Scott is an ESRI-certified ArcView instructor, and is involved in training ArcView and ARC/INFO users on an ongoing basis.

Larry Daniel is president and principal consultant of the Daniel Consulting Group, a GIS start-up venture based in Austin, Texas. Prior to forming this organization, he was vice president of the Castillo Company in Phoenix, Arizona, and director of GIS at MPSI in Tulsa, Oklahoma. Larry holds an M.A. in geography from the University of Texas at Austin and a B.S. in computer engineering from Bucknell University. He has been involved with commercial GIS applications for over eight years. He is a columnist for *Business Geographics*, and has contributed to other professional journals including *GIS World*, *GeoInfo Systems*, *Earth Observation Magazine*, *Computer-Aided Engineering*, and *Design Management*.

Acknowledgments

First and foremost, I wish to acknowledge my debt to Larry Daniel, who initially approached me on this project, and without whom there would be nothing to acknowledge. His vision and expertise are evidenced on every page of this book. My thanks to Gary Irish and Lynn Larson at the Arizona State Land Department for providing the opportunity through our participation in the ArcView Beta program which provided me with the expertise in ArcView to write this book. Thanks to my co-worker, Marleen Riggs, for her valuable perspective as an ArcView end user, and to Jana Fry at Arizona State University, who aspires to be an ArcView power user, and regularly pummeled me with challenging questions toward that end. Special thanks also to Lisa Sivey at GTAC for the insert on cartographic design, as well as being a sounding board for seemingly half-baked ideas.

Thanks also to Environmental Systems Research Institute, Inc., especially Ali Fain, Jack McCarthy, Larry Batten, April Nichols, and Kathleen Bertrand.

And finally, thanks to my mother, who made it all possible–literally– including the computer and the software.

Scott Hutchinson

As second author, I want first to compliment Scott Hutchinson on a thoroughly professional effort in tackling this project. His dedication and perseverance were evident throughout and quite admirable. Apart from Scott, many other parties made sizable contributions and deserve recognition, especially Bruce Hanson who facilitated the use of Equifax/NDS data in the book; Mike Rose, Bill DuBois and Becky Sanchez, who authorized my use of the International House of Pancakes data; and Eduardo Castillo who sanctioned my participation in this project while I was working with the Castillo Company. I would also like to thank many of my associates who are pioneers in the use of "commercial GIS," and who stimulated my thoughts on how to use ESRI's tools in the business world. In particular, this includes Mark Darling and Chris Colas at American Isuzu Motors, Rick Baumgartner at ESRI, and David Huff at the University of Texas. Next, thanks to Richard Turner, Rob Mason, and Jim Sorenson for providing the ArcView application screen shots appearing in our Image Gallery on the companion CD-ROM.

Thanks especially to David Talbott, Barbara Kohl, and the folks at High Mountain Press for providing this exciting technology an opportunity to earn more exposure. And thank you Mellie, my wife, for tolerating the long hours necessary to finish this project.

Larry Daniel

Acknowledgment of Desktop Mapping Innovators

Many organizations and individuals in the private and public sectors have served as innovators in the use and development of GIS and desktop mapping. Several of these innovators' data or ArcView applications appear in this book. Our special innovator list follows: Equifax/NDS; International House of Pancakes; Arizona State Land Department; American Isuzu Motors; Maine Department of Environmental Protection; Dr. James F. Campbell, School of Business Administration, University of Missouri-St. Louis; and Prof. Ben Niemann, Land Resources Department, and Prof. Steve Ventura, Environmental Studies Department, University of Wisconsin-Madison.

OnWord Press Credits

President: Dan Raker
Publisher: Kate Hayward
Associate Publisher: Gary Lange
Acquisitions Editor: David Talbott
Marketing Manager: Janet Leigh Dick
Project Editor: Barbara Kohl
Production Manager: Carol Leyba
Production Assistant: Robert Leyba
Design Editor: Rena Rully
Cover Designer: Lynn Egensteiner
Indexer: Kate Bemis

Contents

Introduction xi
   Why You Should Use This Book xi
   On the Current Status of Desktop Mapping and ArcView . . xii
   How This Book Is Organized xiii
   Typographical Conventions xiv
   Installing the Files from the CD-ROM xv
   On Directories and Operating Systems xix

Chapter 1: Introducing ArcView 1
   The Origins of Desktop Mapping 2
   Desktop Mapping Today 4
   ArcView and Desktop Mapping 7
   The ***INSIDE ArcView*** Perspective 8
   Navigating Windows in ArcView 9
   Windows, Icons and Menus: The ArcView Interface 13

Chapter 2: The Whirlwind Tour 17
   The Sample Project 18
   Exercise 1: A Sample Application 19
      View Navigation Basics *23*
      Cleaning TIGER Street Nets *35*
      Why an Event Theme? *37*

Chapter 3: Getting Started: Projects and Views 47
   The First Step: Opening a Project 48
      Where To Get Data *50*
   Adding Data 52
   More About Themes 54

More About Views 55
Saving Your Work 56
Exercise 2: Opening a Project 57
Importing and Exporting ARC/INFO Data 57

Chapter 4: Extending Data 69
*Defining a Common Ground: Dealing with
Map Projections* 69
Joining Tables 72
Join Versus Link 74
Event Themes 74
The Science of Geocoding 78
Exercise 3: Extending Project Data 81
Primary Research 91

Chapter 5: Displaying Data 93
Defining Symbology 93
Classification 95
Classification Revisited 98
Identifying Features 100
Labeling Features 101
Adding Graphics to a View 102
Exercise 4: Graphics, Symbols and Classification . . . 106

Chapter 6: Data Queries 113
Basic Table Operations 113
Using Tables with Views 115
Selecting Features with Shapes 117
Selecting Features by Query 118
The Undervalued GIS Skill of Database Management . . 119
Locating the Selected Set 120
Logical Queries on Themes 120
A Review of Logical Queries 121
Exercise 5: Making a Smarter Map 124

Chapter 7: Charts 143
Creating a Chart 143
Clarification of Data Markers, Series, and Groups . . . 144
Making Changes to a Chart 146
Using Charts Interactively 148

More on Chart Characteristics 150
Exercise 6: Working With Charts 151

Chapter 8: Layouts 169
What Is a Layout? 169
Why a Layout? 169
How To Make a Layout 170
Layout Frames in More Detail 172
Map Composition 175
On Cartographic Design *176*
Printing 180
Exercise 7: Working with Layouts 180

Chapter 9: Beyond the Basics 189
Overlay Operations 189
Hot Links 193
Working with Shape Files 194
Working with Tables 200
Working with Images and Grids 206
Exercise 8: Shape Files and Hot Links 208
Exercise 9: Working with Images 218

Chapter 10: Optimizing Project Design 223
Data Organization 223
Project Organization 225
Project Optimization 226
Theme and Project Locking 231
Looking Ahead 232

Chapter 11: ArcView Customization 233
Why Customize? 233
Customizing the User Interface 233
Avenue 236
Creating an Avenue Script 240
Communicating with Other Applications 244
Start-up and Shutdown Scripts 245
The Finished Application 246
In Summary 247

Chapter 12: ArcView in the Real World **249**
    ArcView in Government 249
    ArcView in the Business World 253
    ArcView in the Academic World 255
    Elsewhere 259

Glossary **261**

Appendix A: Installation and Configuration **277**
    Memory Considerations 277
    Configuration Considerations 278
    Operating Systems 279

Appendix B: Functionality Quick Reference **281**

Appendix C: About the MicroVision Segments **307**
    The Primary MicroVision Segments for Tempe 307

Index **311**

Introduction

Why You Should Use This Book

Congratulations on selecting ArcView and for choosing this book. ArcView is quickly becoming the premier desktop mapping software package available from Environmental Systems Research Institute, Inc. (ESRI). This book will help validate your organization's good judgment in choosing ArcView.

INSIDE ArcView is not just a rehash of the ArcView manual and on-line help system by writers who are not familiar with the software. This book contains the knowledge that comes from time spent on the front lines of public and private sector organizations in helping individuals, departments, corporations and government agencies learn how to work effectively using geographic information systems (GIS) and desktop mapping software. Our work has included helping business people and government agency professionals who are just getting started with GIS to teaching developers how to customize GIS software to fit their unique environments. We have included tips and tricks that we have learned over time—many of which are either not in the manuals or are difficult to find.

We hope that *INSIDE ArcView* enables you to get "up and running" with desktop GIS. If you have any comments, questions, or recommendations about the book, we encourage you to contact us at OnWord Press, 2530 Camino Entrada, Santa Fe, NM 87505-4835 USA, or e-mail us at readers @hmp.com.

On the Current Status of Desktop Mapping and ArcView

Recent announcements by Lotus and Microsoft of strategic alliances to incorporate desktop mapping functionality in future desktop application suites support the observation that desktop mapping is poised to join the ranks of "mainstream" applications such as spreadsheets, databases, and presentation graphics. Advances in imaging, global positioning systems (GPS) and pen-based computing are further extending the functionality of desktop mapping. Given that an estimated 80 percent of all data has a spatial component, we might ask where desktop mapping does not apply!

All major functionality of geographic information systems (GIS) software is present in ArcView. ArcView has two additional strong points: Avenue, the object oriented programming language with which much of ArcView is written, and direct access to ARC/INFO coverages.

Avenue

The advantages of having access to Avenue for customizing ArcView are covered in Chapter 11, but they are worth summarizing here. ArcView is comprised of objects, and Avenue is an object oriented programming language which allows the user to directly access these objects. Because Avenue provides the ability to access the objects with which ArcView is constructed, it allows a high level of control over all aspects of your environment, including data, applications, interface, and output. Even if you have no immediate intention of customizing ArcView, you can think of Avenue as an insurance policy: when and if you require customization, Avenue's capabilities and power will be available.

Connectivity to ARC/INFO

The ability to directly access ARC/INFO coverages is clearly an advantage to sites using ARC/INFO or with ready access to data in ARC/INFO format. Given that ARC/INFO is the leading UNIX-based GIS software, this is no small advantage. ArcView provides the ability to leverage an existing ARC/INFO installation by adding ArcView Windows or UNIX-based seats to a network. Existing ARC/INFO workspaces can be copied without translation from the UNIX host to stand-alone Windows ArcView installations.

The advantage of access to spatial data in ARC/INFO format extends to users not directly associated with an ARC/INFO site. The widespread use of ARC/INFO among government agencies, coupled with the wide range of exchange formats supported by ARC/INFO, makes it highly probable that most third-party data providers will be running ARC/INFO and as such be able to directly supply data in either ARC/INFO coverage or ArcView shape file format. If spatial data in whatever format is available for your project, it can likely be converted to ArcView format via ARC/INFO.

How This Book Is Organized

ArcView is a program with immense capability and a wealth of features. It can be straightforward and easy to learn if approached correctly. This book is organized in logical and functional sections to help you learn ArcView quickly and efficiently.

Chapter 1 presents a description of ArcView and the historical development of desktop mapping.

Chapter 2, "The Whirlwind Tour," contains a sample project in order to give you the sense of how an ArcView project feels from beginning to end. Data for the sample project and exercises in subsequent chapters are recorded on the companion CD-ROM. In addition, we have saved the end results of the sample project and the exercises in "incremental projects" on the CD-ROM.

Chapters 3 through 8 focus on learning ArcView's basic tools and functionality. Topics include projects and views, data display, data query, extending data, and creating charts, graphs and reports; and layouts (printing custom maps, charts and reports). Exercises provide the opportunity to apply and experiment with what you have learned in respective chapters.

Chapters 9 through 11 focus on advanced topics. Chapter 9, "Beyond the Basics," covers advanced functionality, such as tools for manipulating map projections, hot links to data, and editing tabular data. Three exercises demonstrate these areas.

Chapter 10 covers optimization of project design, and focuses on application-driven projects. Chapter 11, "ArcView Customization," discusses how to customize the interface and provides an introduction to Avenue, the programming language used to write much of ArcView.

The final chapter, "ArcView in the Real World," presents case studies of ArcView projects in government, business and academia which illustrate additional functionality not covered in the sample project that first appeared in Chapter 2.

The glossary contains definitions of selected terms specific to GIS and ArcView.

Three appendices contain information on installation and configuration, a functionality quick reference, and nine MicroVision segments by Equifax National Decision Systems which are used in this book.

Typographical Conventions

✔ **TIP**: *Tips on functionality usage, shortcuts and other information aimed at saving you time appear like this.*

➡ **NOTE**: *Information on features and tasks that is not immediately obvious or intuitive appears in notes.*

✖ **WARNING**: *A handful of warnings appear in this book. They are intended to help you avoid committing yourself to results that you may not have intended.*

The names of ArcView functionality interface items, such as menus, windows, menu items, tool buttons, icons, and dialog box items are capitalized.

> Access the Edit menu, and select Paste. A copy of the theme will appear at the head of the view Table of Contents.

User input, and names for files, directories, variables, fields, themes, tables, coverages, and so on are italicized.

> From the Project window, switch to the *$IAPATH\data* directory, and add the *ihopad2.dbf* and *ihopcmp2.dbf* tables.

Emphasis is indicated by italics.

> In ArcView, the *destination* table is the table to which the fields of the *source table* will be appended.

General function keys appear enclosed in angle brackets. The Shift and Control keys are pressed at the same time as a mouse button or another key. Examples appear below.

<Shift>

<Tab>

<Esc>

<Ctrl>

<Enter>

Key sequences, or instructions to press a key immediately followed by another key, are linked with a plus sign. Examples follow:

<Ctrl>+s

<Shift>+<Tab>

Installing the Files from the CD-ROM

The companion CD-ROM contains data and project files referenced throughout this book, and a gallery of images, or files of ArcView screen captures. There are six directories or levels on the CD-ROM named *avfiles*, *zipfile*, *unifiles*, *vmsv7*, *vmspre7*, and *gallery*.

The *avfiles* level contains the uncompressed data and project files to be copied on the DOS/Windows, Windows NT, and Apple Macintosh platforms.

The *zipfile* level contains a compressed file of the data and project files for DOS/Windows and Windows NT users who would prefer to transfer compressed files to their systems.

The *unifiles* level contains the data and project files to be copied on UNIX platforms.

The *vmsv7* level contains the data and project files to be copied on the version 7 OpenVMS platform.

The *vmspre7* level contains the data and project files to be copied on pre-version 7 OpenVMS platforms.

The *gallery* level contains ten TIFF files of screen captures from sample ArcView applications, courtesy of Environmental Systems Research Institute, Inc. (ESRI). The purpose of the Gallery is to provide you with an "insider's perspective" on the uses of ArcView. The Gallery files can also suggest ideas about how to structure and design your own applications. The sample applications were developed by ESRI, and can be viewed with the use of any graphics application supporting the TIFF format.

As regards mounting and accessing the CD-ROM, some commands are very specific to the type of operating system and/or workstation you are using. Consequently, you should refer to the ArcView CD-ROM installation notes. These notes shipped with your ArcView CD-ROM, and contain the precise commands for mounting the CD-ROM on your system.

Sections on transferring files from the CD-ROM to specific operating systems appear below.

DOS/Windows and Windows NT

ArcView users on these platforms have the options of installing compressed files (a slightly less time-consuming procedure), or uncompressed files. Both are covered below.

Uncompressed Files

1. Create a directory named *insideav* on your hard drive for the data and project files.

    ```
    md {drive:}\insideav
    ```

2. Insert the CD-ROM in your CD drive. If you wish to copy the files from the command system prompt, go to Step 3. If you prefer to copy the files from the File Manager, go to Step 4.

3. To copy the subdirectories and files in the *avfiles* level on the CD-ROM to your *insideav* directory, use the command below. The command assumes the CD-ROM is mounted on the f: drive.

    ```
    xcopy f:\avfiles\*.* {drive:}\insideav /s
    ```

4. To copy the subdirectories and files in the *avfiles* level on the CD-ROM to your *insideav* directory, select the *data* folder on the CD-ROM, click on the Files menu, and select Copy. In the dialog box, type in *{drive:}\insideav*. Follow the same procedure for the remaining three folders on the CD-ROM.

5. In order to access the sample data with the accompanying ArcView project files, you need to set an environment variable on your system. To set this variable, add the following statement to your AUTOEXEC.BAT file:

    ```
    SET IAPATH={drive:}\INSIDEAV
    ```

Compressed Files

1. Create a directory named *insideav* on system.

    ```
    md {drive:}\inside
    ```

2. Insert the CD-ROM in your CD drive.

3. Assuming the CD-ROM has been mounted as the f: drive, copy the file in the *zipfile* level to the *insideav* directory on your system.

    ```
    copy f:\zipfile\av.exe {drive:}\insideav
    ```

4. Issue the following self-extracting command:

    ```
    av.exe -d
    ```

5. In order to access the sample data with the accompanying ArcView project files, you need to set an environment variable on your system. To set this variable, add the following statement to your AUTOEXEC.BAT file:

    ```
    SET IAPATH={drive:}\INSIDEAV
    ```

Apple Macintosh

1. Create a folder on your system and name it *insideav*.
2. Insert the CD-ROM.
3. Double-click on the CD-ROM's icon.
4. Select the *avfiles* folder on the CD-ROM. Inside you will find four folders titled *data*, *projects*, *daycare*, and *work*. Select the four folders and drag them to your *insideav* folder.

UNIX

1. Mount the CD-ROM. Each brand of UNIX workstation has a different syntax. Refer to the ArcView CD-ROM notes.
2. Make a directory named *insideav*.
3. Change your working directory to *insideav*.
4. Transfer the files and directories from the *unifiles* level on the CD-ROM. The following command line assumes that the CD-ROM is mounted as *cdrom*. Substitute your own path for the CD-ROM for */cdrom*. The command line references a shell on the CD-ROM.

   ```
   /cdrom/install.sh
   ```
5. In order to access the sample data with the accompanying ArcView project files, you need to set an environment variable on your system. If you are using the Bourne or the Korn shell, go to step 7. If you are using the C-shell, add the following command to your *.cshrc* file:

   ```
   setenv IAPATH <path to INSIDEAV directory>/insideav
   ```
6. After you have edited the *.chsrc* file, type the following command line:

   ```
   % source .cshrc
   ```
7. If you are using the Bourne or Korn shell, to set the environment variable add the following command to your *.profile* file:

   ```
   $ IAPATH=<path to INSIDEAV directory>/insideav
   $ export IAPATH
   ```
8. After you have edited the *.profile* file, type the following command line:

   ```
   $ .profile
   ```

Open VMS (pre-version 7)

1. Make a directory named *[.INSIDEAV]* on your system.
2. Place the CD-ROM in your CD drive and use the command below:

   ```
   $ MOUNT/OVERRIDE = ID <cd-device>
   ```

3. Install the contents of the CD-ROM level named *VMSPRE7* by typing the following:

   ```
   $ SET DEFAULT <cd-device>:[VMSPRE7]
   $ @INSTALL.COM
   ```

4. Throughout the exercises, the data and project files use a logical to reference their location on your system. To add this logical, edit your LOGIN.COM file in your personal directory, and add the following line:

   ```
   define/translate=(conceal,terminal) IAPATH
   <drive>[<install_directory>.INSIDEAV.]
   ```

5. Execute your login file by typing the following:

   ```
   $ @LOGIN.COM
   ```

6. One last step is necessary to set up files in the *daycare* subdirectory:

   ```
   $ set default $IAPATH:[DAYCARE.]
   $ arc &run import.aml
   ```

OpenVMS (version 7)

Steps 1, 2, 4 and 5 are the same as for OpenVMS (pre-version 7). Substitute Step 3 with the instructions below.

Install the contents of the CD-ROM level named *VMS7* to the *INSIDEAV* directory by typing the following:

```
$ SET DEFAULT <cd-device>:[VMSV7]
```

On Directories and Operating Systems

The data and project files are referenced throughout the exercises in this book as being organized into four subdirectories under a parent directory named *insideav*. The four subdirectories are named *data, projects, daycare,* and *work*. The *daycare* directory contains files and subdirectories of the data and ArcView project files used in the exercise for Chapter 2. The *data* directory contains data files used for exercises in Chapter 3 through 9. The *projects* directory contains ArcView project files used for exercises in Chapters 3 through 9. Throughout the exercises, you are notified about the existence of particular incremental versions of the work sessions. These files can serve as a convenient backup to your own work, or as a means to return to the defaults described in the exercises. Finally, the *work* directory is where you should carry out the exercises in the book.

References to accessing files and directories throughout this book are targeted to the Microsoft Windows environment. Regardless of your operating environment, however, you can perform the exercises in the book.

When you encounter a reference to a directory in the exercises, simply "convert" it to your operating system's syntax. For example, in several exercises you are directed to the *$IAPATH\data* directory. In the UNIX environment, you would convert this directory to *$IAPATH/data*. If you are working in the Mac environment, the directory becomes *insideav:data*. (You would seek the *data* folder in the *insideav* folder.) In the OpenVMS environment, the same directory becomes *$IAPATH:[DATA.]*.

Chapter 1
Introducing ArcView

ArcView is a sophisticated desktop mapping application which promises to bring the power of geographic analysis to the average PC user.

Desktop mapping is built on the concept of spatial data, and spatial data is all around us. An estimated 80% of the data we use has a location component. For instance, it is highly likely that the data maintained by most businesses contain spatial elements, such as addresses, zip codes, or assessor's parcel numbers. With ArcView you can *visualize* site information, customer information, market demographics, and any other type of data with a spatial component.

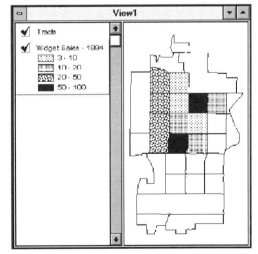

The same data viewed in a spreadsheet and displayed on a map.

The desktop mapping industry is enjoying phenomenal growth, and all signs indicate that this trend will continue well into the future. The following brief outline of the industry's background helps put the phenomenon into perspective.

The Origins of Desktop Mapping

Desktop mapping is one component of a larger branch of information processing called geographic information systems (GIS). GIS arose from the need to incorporate the management of graphic and textual information into a single system. In GIS the linking of graphic information in the form of a digital map, with textual information in the form of a tabular database, produces an "intelligent" or thematic map.

The concept of automated thematic maps is not new. It can be traced to the late 1960s when community planner Dr. Ian McHarg advanced new ideas in the integration of data used in planning decisions. McHarg envisioned a system whereby disparate pieces of data, such as zoning, slope, drainage, and planned communities, could be formed into a cohesive plan via the use of colored acetate overlays. Through visual interpretation of color combinations, optimal sites could be identified for development.

In *Design With Nature*, Dr. McHarg essentially anticipated the emergence of GIS. When addressing the problems inherent in the overlay process for thematic mapping, he wrote, "The mechanical problem of transforming tones of gray into color of equal value is a difficult one, as is their combination. It may be that the computer will resolve this problem, although the state of the art is not yet at this level of competence."

Computer analysts quickly addressed the problems in automating this manual overlay process. Early software systems required that graphics and textual elements be maintained and analyzed separately. The real breakthrough, however, came in the late 1970s with the emergence of software that coupled graphics and textual data into a single system. These integrated systems and the intelligent maps they provided made possible new analytical techniques in which the promise of data synthesis and analysis through GIS began to be realized.

The Origins of Desktop Mapping

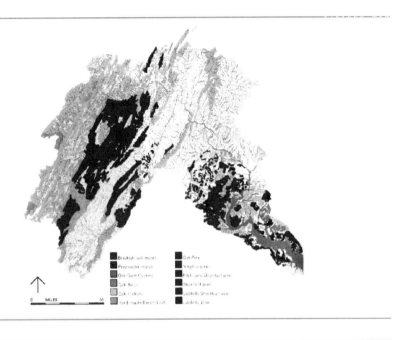

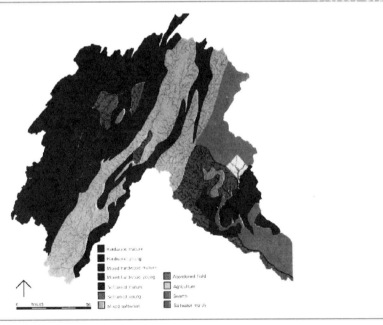

Overlays for wildlife and plant associations from McHarg's Design with Nature (Garden City, NY: Doubleday & Company, Inc., 1971).

The Environmental Systems Research Institute (ESRI) was an early leader in this field. ESRI's ARC/INFO software, released in 1982, quickly became the dominant GIS software. Thousands of ARC/INFO users found at their fingertips a robust and powerful tool kit unequaled in the GIS field.

The advance of GIS in the early 1980s was still hampered by limited (and expensive) computing power, immature software algorithms, and the lack of a user-friendly front end. In addition, data input—a necessity for any GIS project—was difficult and time-consuming. Since the late 1980s, technological breakthroughs resulting in powerful yet affordable personal computers have led to a new class of Windows-based software providing users with the power to perform analysis in a timely manner.

Advances in the availability of GIS data have occurred as well. The expense of custom capture of spatial data, typically by digitizing from paper maps, had previously limited GIS implementation to utilities, government agencies and other large institutions with the resources to subsidize a long-term conversion program. With the 1990 census came the release of TIGER (Topologically Integrated Geographically Encoded Reference) files, arguably the single most significant event in the development in the commercial GIS world. TIGER, and the subsequent release of related data files, represented a wealth of street and demographic data unparalleled in comprehensiveness and format. TIGER became the basis for market research and spawned hundreds of derivative products. Coupled with increasingly powerful hardware and improved graphical interfaces, TIGER has opened the door for smaller government and private users to implement GIS in a cost-effective manner.

Today, GIS is a $3 billion-a-year industry, with an annual growth rate projected at 20%. The hurdles to successful implementation are no longer technical. The only major constraint to increased usage is the ability to visualize how spatial issues affect our world.

Desktop Mapping Today

As mentioned earlier, an estimated 80 percent of the data we use has a location component. By extrapolation, it is not unrealistic to expect at *least* one opportunity every day to produce a map linked to useful data. Appearing below are a few examples of the types of questions that can be answered with the help of desktop mapping software.

A sample of TIGER data labeled with street names and census tracts.

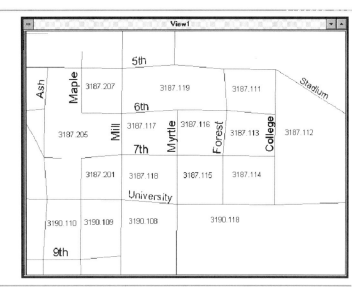

❑ Developers David Z and Zena Y are interested in finding a site for a movie theater complex in the north valley of Seabrook. They notice three suitable properties while driving around in the area. The two decide that they need to examine more comprehensive information on possible sites, including acreage, population growth, and zoning.

David and Zena return to the office and start ArcView. They begin by overlaying a map of current land use in the north valley with a zoning map to identify all vacant commercially-zoned properties. Eight of these land parcels are at least 40 acres in size, prime candidates for a shopping center appropriate for anchoring a theater complex. Next, Zena and David overlay a map showing population growth. Two of the eight parcels are located in areas which have grown by over 20 percent in the last five years. At this point, the two return to their zoning and land use maps, and discover that one of the two parcels is adjacent to a large area of existing and planned multi-family residential units (i.e., sizable apartment complexes). Zena and David make a note to check the ownership of this parcel, and then add a title and legend to the completed map.

❑ A journalist, Harry A, learns from a source close to the Mudville city council that the federal government is planning to relocate the town. Mudville was recently flooded, and the plan specifies a move five miles to the east on federally owned land. Harry wants to know if the projected relocation site is the best possible alternative. He is familiar with desktop mapping, and calls Sally B, a consultant he has used on previous occasions.

When Harry arrives at Sally's office, a scanned aerial photo of the Mudville area has already been loaded into ArcView. Sally overlays the scanned image with a map of the 100-year flood plain. She confirms that the new site for the town is outside the flood plain. But Harry still wants to know if the proposed site would be superior to the current one in the long run. Sally says that something about the aerial photo makes her suspicious. She then overlays a soils map, and discovers that the soils at the proposed site are poorly drained and high in salts. Harry asks, "Is this really the best the federal government can do?"

Sally checks her land use map again, and notices a parcel about a mile farther to the east that is also federally owned. This parcel lacks the soil problems of the current proposed site. While Sally prepares map printouts of both the proposed site and the alternative site with appropriate titles and legends, Harry makes appointments with two federal officials and begins writing the first of a series on the Mudville relocation issue.

❒ The Spicy Abode pizza chain decides to offer home delivery service in Bankersville, a city of 430,000. The company is faced with several daunting challenges, not the least of which is a well-positioned competitor, Pizza Dee-light. Key tactical objectives established by Spicy Abode are the ability to mass merchandise the delivery service in support of all participating pizza franchise stores in the city, and to match the delivery time characteristics to which the market has become accustomed.

Spicy Abode management turns to ArcView and a commercial data provider to define service areas of each franchise store on a city street map. Next, all street addresses within each service area are identified on the map and coded for immediate retrieval.

The Spicy Abode management team decides to offer a single telephone order number for an entire market area. The operators key in pizza delivery customers' addresses in a PC network. Each PC is loaded with ArcView and the city street map arranged by franchise service areas. The operators then direct the orders to appropriate store locations for preparation and delivery.

Our purpose in presenting real life examples was deliberate. Think about the data you work with every day. If you can visualize the data–see the data displayed on a map–you are a prime candidate for desktop mapping.

ArcView and Desktop Mapping

With its ArcView software ESRI has targeted the desktop mapping market with a GIS product designed to answer the need for a layperson's entry into this technology. ArcView firmly lays to rest the impression carried over from early PC-based GIS software that easy-to-use equates with weak. ArcView's many strengths are outlined below.

Power

Built on an object oriented data structure, ArcView's data management and analysis are extremely flexible. ArcView can read data from ARC/INFO coverages (ARC/INFO's spatial data structure), satellite imagery, scanned aerial photographs, dBase and INFO files, external databases such as Oracle and Sybase, and delimited text data. Additional data types can be supported as demand dictates.

The object oriented data structure provides great flexibility with regard to how spatial data can be modeled. For instance, a single ARC/INFO coverage can be modeled as several themes (map layers), each based on a different attribute associated with the coverage. No constraints are placed on how any of these themes are modeled.

Flexibility and Customization

ArcView provides the user with great flexibility in symbolization and map layout. A wide range of symbol and color selection is available, with the ability to import custom symbol sets as needed. The full range of symbols and palettes available for screen display can also be used for generating hardcopy.

The optional Avenue programming language gives the user full control over every element of the ArcView environment. Through Avenue scripts, the user has access to the objects and classes from which ArcView is built. With Avenue, you can customize the interface, customize maps, and imbed macros in your application.

Portability

The ArcView interface is consistent between PC and UNIX platforms. ArcView project files are stored in ASCII text file format, and can be transported between platforms without translation or recompilation. Data are binary compatible between PCs and workstations: an ARC/INFO workspace can be copied from the workstation or PC in native binary format, and used without conversion. This feature positions the PC as a low-cost alternative to UNIX-based development.

Network Savvy

ArcView can access data seamlessly across a heterogeneous network of PCs and workstations. Both ArcView data sets and project files can be distributed across the network. Additional ArcView nodes can be added to the network based on the requirements of the individual user, ensuring access to geographic data as well as other Windows- or UNIX-based applications.

Integration

ArcView links seamlessly with the ARC/INFO GIS software, synergistically extending the power of each. ArcView can read ARC/INFO covers and libraries without translation, and can link to related files in industry standard databases, including Oracle, Informix and dBase.

The INSIDE ArcView Perspective

While ArcView integrates admirably with existing ARC/INFO installations of any scale, its pricing and ability to run on a typical office PC clearly target it to the large group of users who want to perform spatial analysis but have no prior GIS exposure. Included here are many businesses for whom previous GIS software packages were desirable but not cost-effective. With ArcView, the power of GIS and spatial modeling is now within reach.

ArcView is simply a very flexible and dynamic product. Rather than attempt to address all of ArcView's possibilities, we hope to project a specific perspective or vision. Through discussion based on examples, we concentrate on ArcView as a stand-alone PC-based desktop application, and concepts are presented in a manner which is platform independent. Exercises are drawn from a variety of business and government applications.

Note that our focus does not mean we will be confining ourselves to the basics. Many tasks of seemingly great complexity can be accomplished with ease—if you first know where you are heading. In Chapter 2, we focus on a sample project from beginning to end as a means to help you navigate through the ArcView tool kit.

Before we begin, a few housekeeping chores are in order. If you are already experienced in working with a windows-type graphical user interface (GUI), skip the next section and proceed to "Windows, Icons, and Menus: The ArcView Interface."

Navigating Windows in ArcView

ArcView runs in five different operating environments. In all these environments, the graphical user interface (GUI) is comprised of windows and diverse types of controls. Throughout this book we assume that you have a basic familiarity with your windowing environment. For those who are new to working with a GUI, the following description of basic windowing terminology may prove useful. For those requiring additional orientation in the use of the mouse and windows, access the windows tutorial available on your system or your system documentation.

> ✒ **NOTE**: *This book is oriented toward the Microsoft Windows version of ArcView. Users of ArcView on other platforms will find that while all text and exercises are applicable, certain windowing and operating-system dependent functionality may differ. Included here are*

the particular mouse button used for certain functions, and navigation of the underlying file system.

Clicking

Clicking refers to pressing and releasing a mouse button. Typically, clicking is used to select a program feature, access a menu, or activate a tool or button icon. In Microsoft Windows, all clicking is done with the left mouse button.

Double-clicking

Double-clicking refers to quickly pressing and releasing a mouse button twice in succession. Double-clicking is commonly used to select a program feature or activate a function associated with that feature.

Mouse Button Functions

Responses associated with mouse button actions—clicking, double-clicking, selecting, and dragging—are collectively referred to as mouse button functions. These functions are a combination of the operating environment and the application software.

Dragging

Dragging refers to moving the mouse pointer to an object, pressing and holding down the mouse button, moving the pointer to a new location, and releasing the mouse button. Dragging is commonly used to resize or reposition objects on the screen.

Mouse Pointer

The mouse pointer is a symbol on the screen which accompanies the movement of the mouse. Common pointer symbols are an arrow and what appears to be a large capital I.

Menu

A menu is a list of application functions. Typically, menu titles appear in a horizontal bar at the top of your screen. The bar containing the menu titles is called a *menu bar*. To activate a list of menu options, move the mouse pointer to the menu title, and then click the left mouse button. Another method of activating a list of menu options is to press a function key (e.g., <Alt>, <Ctrl>, <Shift>) followed by a character key. The character key used to activate a particular menu may be indicated by an underlined character in the menu title.

Icon

Icons are graphical image shorthand for program functions. Icons (little pictures) appear on buttons or by themselves. Clicking on an icon or a button containing an icon activates a particular program function or operation.

Pull-down Menu

When you click on a menu title in a menu bar, a box containing a list of menu functions and options appears under the menu title. The box is called a pull-down menu. Pulling down a menu is somewhat similar to pulling down a window shade.

Status Bar

The status bar is a narrow bar at the bottom of your screen. Its purpose is to provide cues on program functions, as well as information during program processing. In many applications moving the mouse pointer to a button containing an icon will trigger the appearance of a brief description of the button's function in the status bar. Next, the status bar will inform you of the progress of an operation that you have executed (e.g., *30% complete, 60% complete*).

Dialog Window

Dialog windows appear when you activate a menu option or a button, and the program requires additional input. Assume you want to print a page. First, you would activate the print menu. At this point you may see a dialog window requesting additional information, such as the number of copies, and the number of the page you want to print. You can choose the defaults, or input your choices via the keyboard.

Resizing

When dragging is initiated upon placing the pointer on the side or corner of a window, the window size will be adjusted to the new dimensions. This is referred to as resizing the window.

Repositioning

When dragging is initiated by placing the pointer in a window title bar, the window will be moved to the new location. This is referred to as repositioning the window.

Make Active

When multiple windows are present on your screen, work is focused on the window which is active. To make a window active, move the mouse pointer anywhere within the window and click the left mouse button.

Maximize

To maximize a window, click on the Maximize button in the upper right hand area of the window. The window will be resized to fill the entire screen.

Minimize (Iconify)

To minimize or iconify a window, click on the Minimize button in the upper right hand area of the window. This will collapse the window so that it is represented on the screen as an icon.

Restore

Double-clicking on an iconified (minimized) window will restore the window to its previous size.

Bring to Front or Forefront

If multiple windows are present on the screen, clicking in the title bar of a window which is partially covered will make that window active. In the process, the window will be brought to the forefront of the screen (to the front of the other windows). If the window is completely covered, simultaneously depressing the <Alt> and <Esc> keys will bring the covered window to the front of the screen.

Windows, Icons and Menus: The ArcView Interface

ArcView organizes your mapping project and the tools available to you within a system of windows, menu bars, button bars, tool bars and icons. The entire ArcView environment and GUI are contained in the main application window. All user interactions take place in this area, as well as the display of ArcView output.

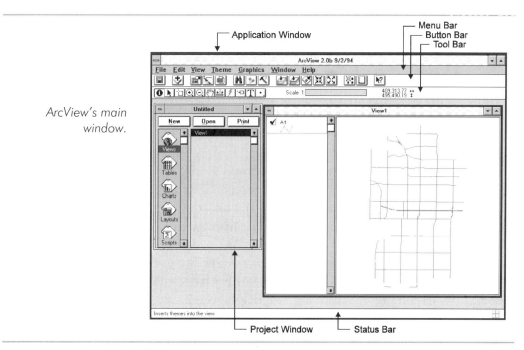

ArcView's main window.

Additional ArcView window elements.

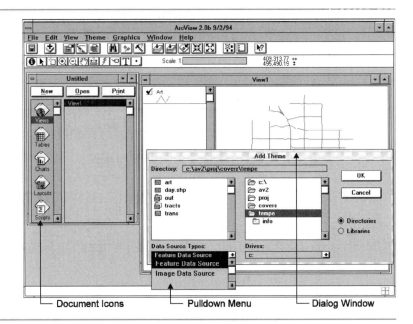

There are several ways to interact with ArcView. Pulldown menus are the most conventional method. Choosing menu items communicates commands to ArcView by executing embedded scripts (macros or mini-programs). Menus can be pulled down and selected in one mouse operation, or pulled down with one mouse click and selected with a second.

Actions are also conveyed with buttons and icons. A single mouse click activates the button bar and tool bar icons. Clicking on an icon may produce an immediate action or bring forth an additional input window, depending on the tool. These additional input windows are known as *dialog windows*.

A helpful feature in navigating the icons is the *status bar* at the bottom of the application window. When the cursor is placed over an icon, a brief description of the icon's function is displayed in the status bar. For example, if the cursor is placed over the Add Theme icon (the one containing a plus sign), "Inserts New Themes Into The View" is displayed in the status bar. The status bar can be invaluable to new users and to those who find icon symbols less than intuitive.

ArcView presents all of its features in a series of nested windows. The top level or main window is the *project window*. The project window organizes all files and documents that you generate, access, and input when using ArcView, including spatial data, associated tabular data, scanned images, charts, graphics, and scripts. An ArcView project organizes these documents in multiple related windows, and these windows are all accessible from the project window. ArcView projects can be saved, and when a saved project is reopened, the project is returned to the exact state of the last save, with all windows and associated graphics and tables displayed as before.

There are five document categories used in ArcView: *views, tables, charts, layouts* and *scripts*. Each document category contains a list of available documents created as part of the current project. Each document (e.g., a specific view or table) can be opened by double-clicking on the name of the document from the document list, or by selecting the document name and then clicking the Open button. Other documents and files associated with each document will be displayed in separate windows. For example, in addition to a theme displayed in a view window, a table and a chart associated with the theme may be displayed in other windows.

It is important to note that the tools available on the menu bar, button bar and tool bar will change depending on which document type is active. The tools available for working with views are different from tools available for tables, as are the tools for charts, layouts, and scripts.

Do not feel intimidated if this organization sounds foreign or unfamiliar. Even experienced GIS professionals are bound to spend some time learning the vernacular. Entire chapters are devoted to ArcView projects, views, tables, charts, and layouts. Scripts are addressed in Chapter 11. For the moment, just be sure you have this section marked because there will be frequent references to the above terms.

Chapter 2

The Whirlwind Tour

A central concept of GIS is viewing your data spatially. For many people trained in the "flat" world of spreadsheets and databases, the spatial mindset required in GIS is not at all intuitive.

GIS requires users to think more holistically than other software applications. In spreadsheets and databases, you can enter data, and postpone decisions about how to conduct analysis without worrying about whether you will ultimately be able to obtain the answers you seek. But GIS is different. There are countless possibilities for how data might be organized and modeled. Unless you know where you are going from the outset, you may never complete your projects effectively.

In this chapter we will use a sample project to show how ArcView is applied to this process. A project is the overall structure used in ArcView to organize the files of your work session, such as digitized maps, tabular data and presentation graphics. We will introduce you to the "look and feel" of ArcView, and you will see how an ArcView project feels from beginning to end. In subsequent chapters, the program topics and functions covered below are discussed in detail.

As you thumb through this book, you will notice that we have interspersed exercises throughout the chapters. Most of the exercises are short because they have been tailored to illustrate specific concepts. The sample project exercise in this chapter is lengthy because we will be illustrating the whole. Follow along with us, and if you get lost, do not worry. We have saved incremental versions of the sample project on the companion CD-ROM so that you can always check to see where you should be at any point along the way. (If you installed the *projects* directory from the CD on your hard disk, you can access the incremental projects from your system.)

The Sample Project

In this example, Sharon Y seeks to open a new day care facility. Where would she ordinarily start? If she is like most small business people, she sells her general concept to investors first and then contracts with a commercial real estate broker to locate available properties which match her needs. The basic flaw in this arrangement is that the entrepreneur and the investors are postponing the most important aspects of their venture (location, location, location).

An alternative arrangement would be for Sharon to carry out locational analysis at the outset, that is, identify the ideal location, inquire about likely cost, and then ask her investors for the necessary funding. This is a win-win situation for everyone: Sharon is better prepared to acquire a winning location, and her investors can be more confident about supporting a successful venture. One could argue that Sharon's "optimal" location may not be for sale. Many professionals' experience, however, indicates that real estate can be acquired from property owners who are not currently in the market if they are approached by an interested and motivated buyer. Thus, in our hypothetical world, Sharon kicks off her venture with locational analysis.

The first step in Sharon's new venture would be to identify the most promising locations in her market area. In the day care business, an entrepreneur may wish to focus on the following variables:

- Population distribution
- Number of children age six years or less as a proportion of the total population
- Average household income

By classifying various neighborhoods according to how they compare on these criteria, ArcView can be used to visually portray how promising various locales may be for a new day care center. Next, Sharon can also begin to identify competitors' market areas by adding existing day care facilities to a map created in ArcView.

Subsequent steps could involve examining the most promising locations more closely. For example, once the primary criteria have been explored, Sharon can study secondary criteria such as the number of working mothers with children under age six as a proportion of the total population, local occupation distribution, educational attainment, population projections and the locations of major employers.

With GIS mapping as support, Sharon will be much more effectively positioned to initiate her property search. Instead of analyzing the suitability of available properties, she can direct her energy toward analyzing the availability of suitable properties.

Throughout the remainder of this chapter, we will act as Sharon's GIS support by using ArcView to help her identify suitable sites for her business. We could tell you how to use ArcView, but GIS makes more sense–and it's a lot more fun–when you get in there and do it yourself. We encourage you to perform the analysis along with us.

Exercise 1: A Sample Application

Our application is straightforward. We need to identify the most promising locations to open a new day care center in Tempe, Arizona. We will prepare a map locating Sharon's potential customers and the existing competition. Our data sets include demographic data from the 1990 census, with total population by census tract as well as population age six or less; Yellow Pages listings for day care centers with street addresses; and spatial data sets for Tempe.

The spatial data sets are derived from the U.S. Census Bureau's TIGER files, and consist of (1) a file representing census tracts, called a *coverage*, and (2) a street map (*street net*) of Tempe which has been digitized. The street net file contains codes that represent address ranges (e.g., 1400 N. 15th Ave. to 1500 N. 15th Ave.). Both spatial data sets include tabular data consisting of codes that represent geographic locations. (See the insert, "Cleaning TIGER Street Nets.")

> **NOTE**: *In the following exercise, the spatial data sets (the Tempe census tracts and street net) are stored separately from the non-spatial or tabular data (the demographic data and competing day care locations). The two types of data files will be subsequently linked together using a locational field, such as a census tract number or a street address. This separation of spatial data and attribute data is typical of most GIS projects.*

20   Chapter 2: The Whirlwind Tour

Begin by starting ArcView. Drag the lower right corner of the application window to fill the entire screen.

Our first task is to prepare ArcView for importing data files into the project.

1. Click on the Views icon, and select New. A view window is opened, with the default title of *View1*.

2. Drag the lower right corner of the view window to fill the available space in the project window. Because the *View1* window will display graphics and analysis results, it should be as large as possible.

Before proceding with the exercise, we need to set the working directory for the project. Click on the project window (labeled *Untitled*) to make it the active window. If necessary, move or close the view window to make the project window visible. From the Project menu, select Properties. The default setting for *Work Directory* is set as *$HOME*. Change the name to *$IAPATH\work*, and click OK to accept the change. If you previously closed the view window, click again on the view window (labeled *View1*) to make it active.

NOTE: *The $IAPATH variable references the directory where you installed the INSIDE ArcView sample data. If you have not set the $IAPATH variable in your autoexec.bat file, refer now to the instructions in the Introduction on installing the sample data from the CD-ROM.*

You are now ready to add the spatial data sets, or Tempe geography in this case. In ARC/INFO, these data sets are referred to as *coverages*. In ArcView, spatial data sets–coverages and shape files–are referenced by *themes*. When spatial data is imported into ArcView, it is transformed into a theme. Think of themes as maps and map overlays. We will now procede to import four themes into ArcView.

1. On the button bar, click on the Add Theme icon.

The Add Theme icon.

Exercise 1: A Sample Application   21

A dialog window is opened for additional input.

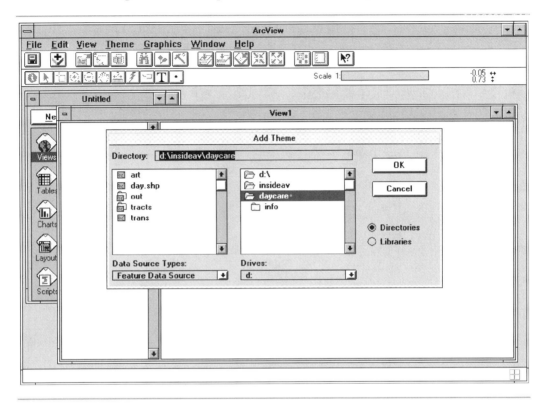

Adding themes to the view.

2. In the Add Theme dialog window, click on the Directory box until you see the day care directory installed from the CD-ROM (*$IA-PATH\daycare*). Four themes are listed: *art* (arterial street net for Tempe), *out* (outer boundary of Tempe), *tracts* (census tracts for Tempe), and *trans* (street net for Tempe), as well as one shape file, *day.shp*. Select the four themes.

✔ ***TIP***: *After clicking on the first selection, hold down the <Shift> key while clicking on additional selections. If you do not hold the <Shift> key down, each theme selected replaces the previously selected theme.*

22 *Chapter 2: The Whirlwind Tour*

3. After all four themes are selected, click on OK. The dialog box is cleared, and the themes are added to the view. A view can be conceptualized as a collection of map overlays (themes).

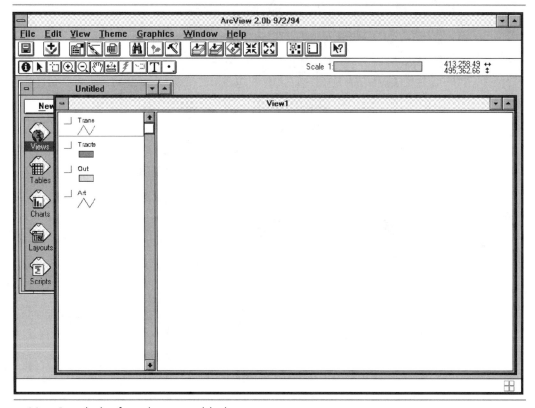

View1 with the four themes added.

Let's take a peek at the raw data by clicking on the check box adjacent to the *tracts* theme. The census tracts for Tempe are drawn in the graphics display area of the view window. By default, the theme is drawn in uniformly shaded areas. The demographic data we intend to add later will be displayed over the census tract geography.

Exercise 1: A Sample Application 23

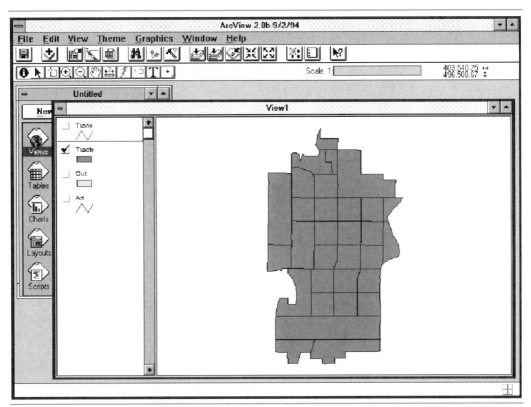

The census tracts theme for Tempe.

View Navigation Basics

The first thing you may wish to do upon displaying a new theme in a view is to take a closer look. Nestled in the View graphical user interface are the tools to do just that—zoom in, zoom out, and pan from one area to another.

The Zoom In and Zoom Out icons from the View button bar.

The Zoom In and Zoom Out buttons on the View button bar provide the basic functionality of zooming in and out in a view. Clicking on these buttons will cause the display to be zoomed in or out by a factor of two, centered on the extent of the original display. These buttons provide a fast means of changing the extent of a view.

The Zoom In, Zoom Out and Pan icons on the View tool bar.

The Zoom In and Zoom Out tools from the View tool bar provide an expanded set of functionality on the Zoom In and Zoom Out buttons described above. Instead of immediately zooming in or out, the Zoom In and Zoom Out tools become active when clicked on, with the result dependent on the action subsequently taken in the view. Clicking on the Zoom In tool to make it active, followed by clicking on a point on the display will cause the view to be zoomed in by a factor of two, centered at that point. Alternatively, a box can be dragged using the mouse, and the display will be redrawn containing the extent described by the box. Zooming Out functions by clicking or dragging a box on the view in much the same manner.

The Pan tool is used to change the extent of the display without zooming in or out. Clicking on the Pan tool changes the cursor to resemble a hand. Clicking on a point in the view and dragging the display while holding the mouse button down causes the map to be dragged to a new location. Releasing the mouse button results in the display being redrawn with this new extent.

The Zoom to Full Extent and Zoom to Active Themes icons from the View button bar.

It is often desirable to easily return to the full extent of the map after zooming in to a selected area. The Zoom to Full Extent button will redraw the display zoomed to the extent of all themes in the view, or the full map extent of the view. Alternatively, the Zoom to Active Themes button is used to zoom to

> the full extent of only those themes currently *active* in the view. Clicking on the entry for the theme in the Table of Contents causes the theme to become active. The theme is highlighted in the Table of Contents, and appears as a raised box. If the extent of this active theme is less than the full extent of all themes in the view, clicking on the Zoom to Active Themes button will cause the display to be zoomed to the extent of the active theme. If more than one theme is active, clicking on the Zoom to Active Themes button will cause the display to be zoomed to the combined extent of the active themes. (To make more than one theme active, hold down the <Shift> key while clicking on additional themes in the Table of Contents.)

Next, we need to import the tabular data sets. These tables contain our demographic data and locations of competitor day care centers. In preparation for this exercise, the tabular data were entered in ASCII text files with the fields separated by commas. (Through this format, you should be able to import data from almost any source. ArcView can also import data directly from dBase format (*.dbf*), and from INFO files, the default database supplied with the workstation version of ARC/INFO.)

✔ *TIP: When ArcView imports a comma-delimited ASCII file to its tabular format, the program searches the first line of the text file for the field names. Thus, the first line of the text file should contain the names for each field separated by commas. Field names can include spaces, and need not be preceded and followed by double quotation marks.*

The data in the fields are recorded in all lines but the first line of the text file. The fields in the data lines (records) are separated by commas. A field beginning with a numeric value which must be treated as a character string (such as an address field) should be enclosed in double quotes. Finally, the file must be given a .txt extension in order for it to be listed in ArcView's Import menu.

To import the tabular data, take the following steps:

26   Chapter 2: The Whirlwind Tour

1. Click on the Tables icon in the project window. Note that the menus and icons displayed in the application window change. Before you add any tables to your project, you are presented with few choices.

2. Pull down the Project menu from the menu bar, and select Add Table. As seen in the following illustration, an Add Table dialog box is opened.

3. Navigate to the $IAPATH\daycare directory.

4. Pull down the List Files of Type menu, and select Delimited Text (*.txt). Two files are listed: *day.txt* and *maricopa.txt*. Select both, and click on OK.

The text files are imported and displayed in tabular format. You can stretch and scroll these windows as you wish to examine the data.

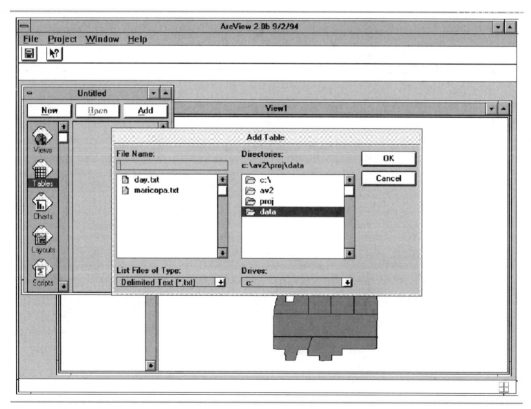

Adding tables into View1.

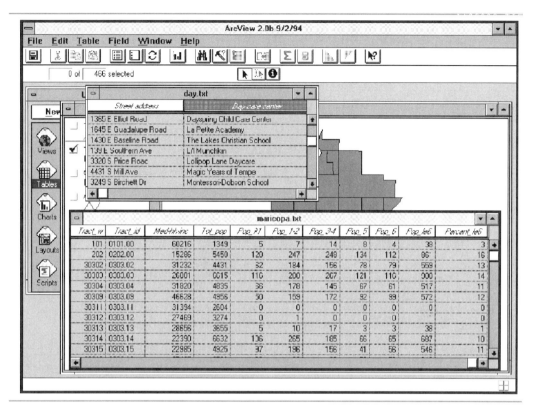

A closer look at the tables.

5. The *day.txt* table can be closed, because we will not need it until later. Return to the spatial themes by clicking on the Views icon from the project window, and opening *View1*. (For your convenience, we have saved incremental versions of the exercise project files on the CD. These incremental projects serve as a reference point in case you need to return to our defaults. The incremental project to this point has been saved as *ch2a.apr*.)

At this point we have imported all of the raw data into ArcView. Now we need to link data together. The first step is to *join* the two tabular data files to the two spatial data themes so that the tabular data can be spatially displayed. We will begin with the demographic data in order to display it against the census tract theme.

28 Chapter 2: The Whirlwind Tour

Tabular data can be linked to a theme via a field which identifies the geographic location. Take the following steps to join the demographic tabular data to the census tract theme.

1. Click on the *tracts* theme in the legend of the *View1* window in order to make it the active theme. (When active, the legend for the theme appears raised in the legend area of the view window.)

2. Select the Open Theme Table icon from the button bar. (Refer to the text displayed in the status bar to confirm you are making the proper selection.)

The Open Theme Table icon.

The following illustration shows that the attribute table is opened for this theme, displayed as *Attributes of Tracts* in a new window.

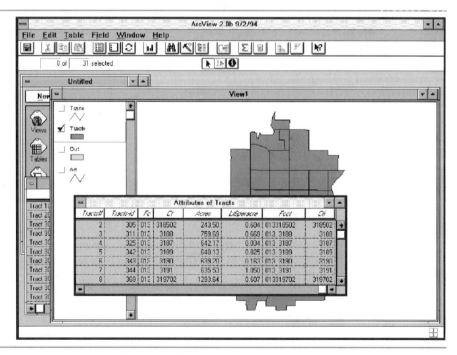

The opened Attributes of Tracts table.

Exercise 1: A Sample Application 29

3. Bring the window displaying the *maricopa.txt* table to the foreground, and stretch and position it so that this window and the Attributes of Tracts window are displayed simultaneously.

4. To join two tables, we need a common field containing the same values in both tables. For a one-to-one association to be made, these values must be unique in both tables. In our example, the unique field held in common between the two tables is the census tract number. In the *Attributes of Tracts* table, this field is named *Cti*, and in the *maricopa.txt* table, it is named *Tract_w*. Click on the name of this field to be used to join the two files in the column heading of the file. The name will be shown as highlighted when selected.

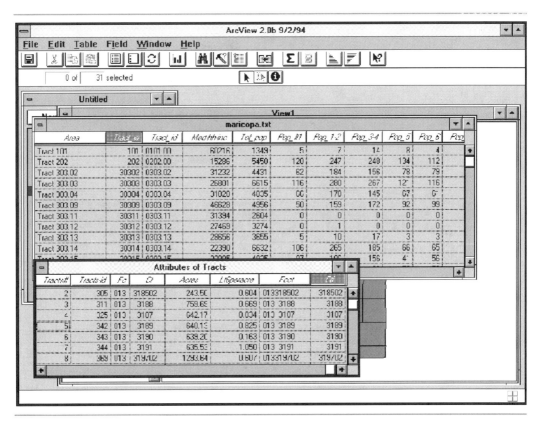

The Attributes of Tracts and maricopa.txt tables prepared for joining.

5. Click on the title bar for the *Attributes of Tracts* window to make this table the primary table for the join. (When joining tables, the attribute table for the spatial theme should always be the primary table.) Pull down the Table menu from the menu bar and select Join; the status bar of the application window will show progress of the operation until the join is complete.

When the join is complete, the *maricopa.txt* file will be closed, and the *Attributes of Tracts* table will contain the fields from the *maricopa.txt* table columns to the right of the spatial theme columns. You can scroll through the table to examine the results of the join.

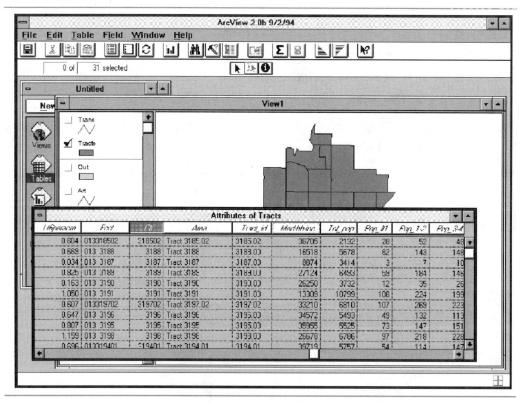

The resulting joined Attributes of Tracts table.

At this juncture, the demographic data is associated with the census tract polygons. We can now make a *thematic map*. A thematic map displays a set of related geographic features. Typically, a *classification* scheme has

Exercise 1: A Sample Application   31

been applied so that the map displays non-spatial data associated with the geographic features.

1. Close the window displaying *Attributes of Tracts*, and click on the Views icon. Open *View1*, and the legend of available themes is displayed.

2. Now we can classify the *Tracts* theme. Double-click on the *Tracts* entry in the *View1* legend to bring up the Legend Editor. Click on Labels, and pull down the Field menu.

3. For the field to classify, select *Pop_le6* (the number of persons age six or under by census tract). When the default classification is complete, five quantile classes are displayed as illustrated in the following figure. By default, the classes are shaded in a graduated gray scale.

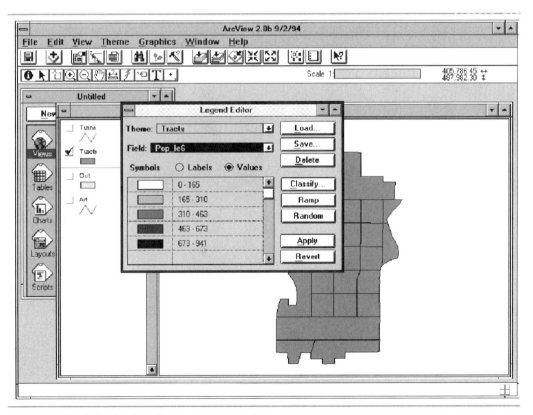

The default classification on Pop_le6.

32 *Chapter 2: The Whirlwind Tour*

4. In order to more clearly show the breaks in the data, you can change the initial color assignments by double-clicking on the first symbol to bring forth the Symbol Editor. From the Symbol Editor, you can change properties such as outline width, fill pattern, and color. Experiment with color choices until you have a satisfactory combination, and then click on the Apply button in the Legend Editor to apply these changes to the map.

5. The screen is redrawn with your new choices. If you do not like the final result, try again. If you like your results, go ahead and Save them now. We have saved our choices for you to fall back on. (The incremental project has been saved as *ch2b.apr*.)

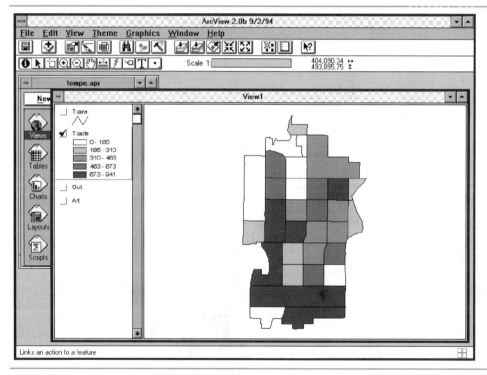

The applied classification on the Tracts theme.

The demographic data has been imported, joined to the *Tracts* theme, and classified. Our initial work aimed at locating concentrations of children under age six is complete. Now we can proceed with the data on competitors.

Exercise 1: A Sample Application   33

We intend to locate competitors by matching their street addresses to a street network (map). This process of address matching is referred to as *geocoding*. One of the themes you initially imported was the street net theme for Tempe, illustrated in the figure below. Although this street net theme is coded with the attributes necessary for address matching in ArcView (i.e., street names and address ranges), the theme must first be made *matchable*. The theme must be matchable before you can geocode a data file containing street addresses against it.

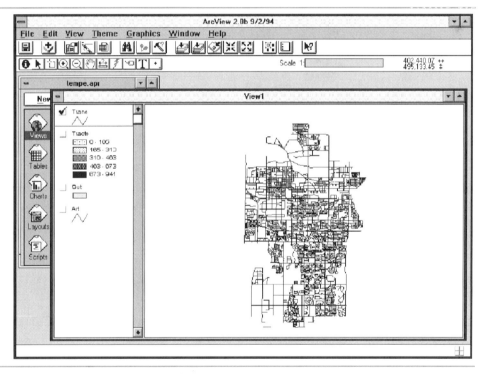

The street net theme (Trans) for Tempe.

To make the theme matchable, take the following steps:

1. Make the *Trans* theme active by clicking on its entry in the legend area of the *View1* window.

2. In the Theme pull-down menu from the menu bar, select Theme Properties.

34 Chapter 2: The Whirlwind Tour

3. You will see a window containing a number of theme operations. Scroll down the icons on the left side of the menu until you find a mailbox icon labeled Geocoding. Clicking on this icon brings forth another dialog box asking you to identify the fields to be used in address matching.

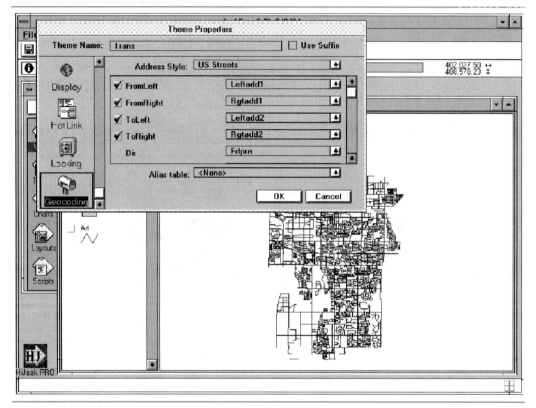

The Theme Properties dialog box.

4. Because we are using a standard TIGER street net, ArcView recognizes the standard field names present. Accept the default choices, as well as the default Address Style, US Streets. Upon clicking the OK button, the software processes the theme. The theme is now matchable.

Cleaning TIGER Street Nets

In Exercise 1 when the address data is geocoded to the street net, we obtain a 100% match the first time around. While there is no reason to expect a nightmare when you attempt this on your own, the variability in address formatting and the dynamic nature of an urban street net dictate that a few hiccups on the way to a successful match are to be expected. Errors in the street net and in your data file are effectively inherent to the geocoding process.

The most readily available street net for the United States is the TIGER street net, available from the U.S. Census Bureau. Priced at approximately the cost of distribution, TIGER is a bargain, but it is not without problems.

The main problem with TIGER is basically one of completeness. TIGER was created for use in the 1990 census data collection effort. The cutoff point for entering new street segments occurred prior to 1990. If you are working in a rapid growth area, you can be sure that many street segments will be missing. In addition, the existing street segments may be coded with incorrect address ranges, or they may lack address ranges.

In the course of preparing Exercise 1, five of the 23 street addresses did not match (were not located) on our first attempt. Upon examining the TIGER street net for Tempe, we found one missing street, which we digitized and coded manually, and four streets with missing or incorrect address ranges, which we fixed manually. We did not search for nor fix any additional coding errors in the street net. You can verify this by examining the attribute table for the entire street net.

A host of vendors have come forth to provide better and cleaner TIGERs. The vendors have fixed the coding errors, added the missing address ranges, and in many cases digitized additional street segments not present when TIGER was created. You will pay a little more, but it may well be worth it.

> The second source of error is in the address field of the database you are attempting to geocode. While a full discussion of geocoding and address formats takes place in subsequent chapters, suffice it to say at this juncture that spelling errors in street names, incorrect street types, and inclusion of additional data in the primary address field, such as apartment or suite numbers, can serve to reduce the accuracy of matching. You can use ArcView's edit window to correct these errors on the fly, but this is no substitute for having your data clean and properly formatted at the outset.

We are now prepared to match the addresses from *day.txt*, the table of competing day care centers. The first step in this task is to create an *event theme* from this table. (See the insert titled, "Why an Event Theme?")

To create an event theme, take the following steps:

1. Click on the Views icon from the project window and open *View1*.

2. From the pull-down menu on View, select Add Event Theme.

3. A dialog box opens asking us to identify the event theme type. From the three icons in the upper left corner, select the far right icon, Address Theme.

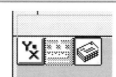

The Add Event Themes icons.

4. We are then asked to identify the existing address theme (the theme to be made matchable); respond with *Trans*.

5. Leave the Join Field blank, because we elect not to physically join the two tables. Next, specify *day.txt*, the name of the table from which to create the event theme.

6. From the pull-down menu associated with the Address field, select the Street Address field. Accept the default offset distance of zero, and click on the OK button to proceed.

➡ **NOTE:** *For address matching, the entire address should be contained in a single field.*

Why an Event Theme?

In the sample project appearing in Exercise 1, we work with both spatial data sets (census tracts, street net) and tabular data sets (demographic data, competitors' addresses). In working with the demographic data, the link between the spatial and tabular data sets seemed clear: we joined the two tables on a common attribute, and as a result of this join, we were able to display the demographic data spatially, that is, map it by census tract.

However, when preparing to work with address data and join it to the street net, we discover that the relationship is not so clear. Before this link can be performed, an "event theme" must be created from the address data. What's so different about addresses? For the answer to this question, we need to look under the hood briefly and examine how data is stored and modeled in ArcView.

The original data structure beneath ARC/INFO, and carried forward into ArcView, is a georelational data model. In this model the spatial data—the lines, points and polygons which are used to represent the real world—is stored in one set of files, and the attribute data associated with these features is stored in another set of files. The attribute data is linked to the spatial data in a one-to-one relationship. The *relate* concept is paramount: while the data model can be extended through multiple relates, one-to-many relates, external relational databases and the like, it is still built upon, and limited by, the foundation of relational database technology.

In ArcView, we find new software built from the ground up using an object oriented data model. As a result, ArcView is much more flexible about data types and linkages than ARC/INFO, but the underlying framework remains. ArcView still "likes" to see a one-to-one relationship between spatial features and the attributes associated with these features.

Address geocoding is too unwieldy to maintain a discrete node in a street net for every address. Instead, each street segment is coded with address ranges, and the software interpolates along the segment to locate a specific address. These interpolated points are determined on the fly as the data file containing street addresses is geocoded against the street net.

38 Chapter 2: The Whirlwind Tour

In the ArcView world, however, we want to preserve the results of this linkage for future analysis. To make the outcome of on-the-fly interpolation more lasting, ArcView creates a *shape file*, a new spatial data set comprised of the point locations for each address located, as well as the link back to the address record in the data file. The shape file is not an ARC/INFO coverage, but in ArcView, it works virtually the same way. The outcome is yet another instance of our basic data model (a spatial data set, or series of geographically distributed points) linked to an attribute table (address records and other associated data). And like any other ArcView theme, we can display, classify, and model it however we wish.

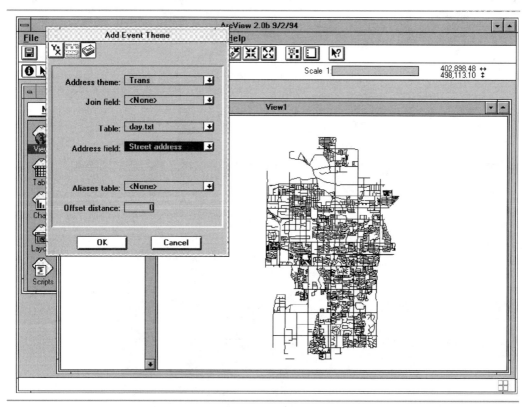

The Add Event Theme dialog box.

Exercise 1: A Sample Application 39

7. We are now prompted via yet another dialog box to enter the Geocoded Theme Name. This is the name and location for a new type of theme, called a *shape* theme, that we will create at the completion of geocoding the data table. ArcView is going to create the new theme containing the point locations for all the locatable addresses in the data file. Save this theme in your current working directory—not the *daycare* directory—with the title *day.shp*.

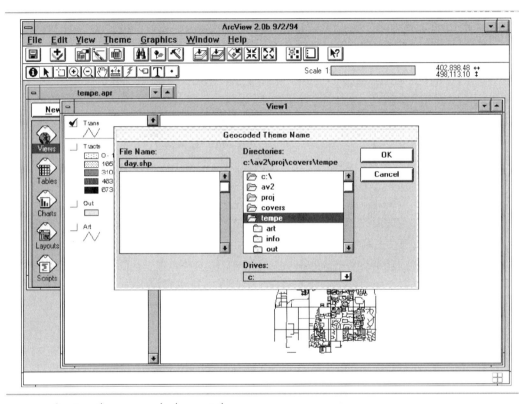

Saving the geocoded event theme.

8. Having entered and given the OK on the theme name, the address match is set to start. At this point, the Geocoding Editor window appears.

40   Chapter 2: The Whirlwind Tour

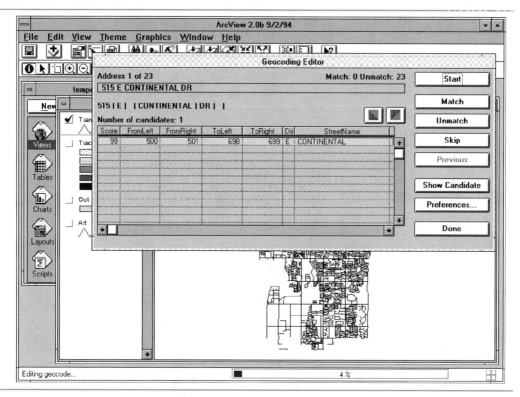

The Geocoding Editor dialog window.

At the start of geocoding, the first record from the data file is presented along with all possible street segment candidates in which to locate this address. In addition, a score is given to rank the completeness of the match to the street segment from the street theme. (By default, the threshold for determining a successful match is a score of 50.) If there are multiple candidates, they are listed by descending score value, with the best match listed first. If there is no match, the field is blank. If you are lucky, there is one candidate shown, for which the matching score is high. In our example, the first matching record is *515 E. Continental Dr.* with a score of 99, for which we are given one matching candidate street segment. This is most definitely a match.

9. At this point we can proceed in two ways: step through the data file record by record to evaluate all matches, or let the software match as many as it can without user interaction. The success and high score of the first match assure us that the address field is properly formatted. Rather than step through record by record (using the Match button), we elect to match the entire file. Select the Start button.

10. In our sample application, all 23 records match. Accept the results of the geocoding process by clicking the Done button.

11. At this point the *address matched* event theme is added to the view. This theme is like any other in that it can be viewed, queried, and classified. Use the Symbol Editor to change the color of the dots if you want more contrast. (The incremental project has been saved as *ch2c.apr*.)

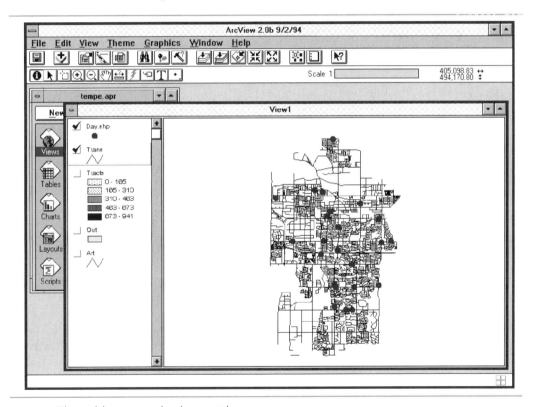

The address matched event theme.

We are now ready to display the locations of existing day care sites against the five classes of population age six and under. Click on the check

42 Chapter 2: The Whirlwind Tour

box to display the population theme alongside the day care sites, as seen in the next illustration.

The drawing order of themes in ArcView is controlled by their positions in the legend. ArcView themes draw from bottom to top. To change the drawing order of a theme, click on the theme in the legend while holding the mouse button down, drag the theme to a new position in the legend, and release the mouse when you have it where you want it. The display will redraw automatically.

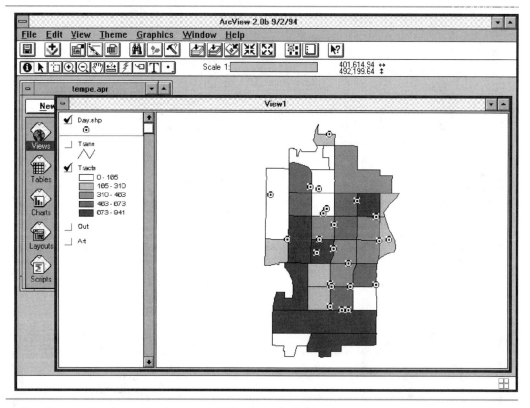

Displaying the day care theme with the Pop_le6 theme.

The darkest areas are those with the largest number of kids age six and under. The location of existing day care sites suggests several promising areas for the location of a new center. But, we are not yet ready to make recommendations to Sharon Y. The demographic numbers are raw counts, and the census tracts are not all the same size. While the raw counts are useful because Sharon needs a minimum service base regardless of population density, we still need to know more.

Another field in the demographic data set, percentage of total population age six and under, might be useful here. While this field does not normalize the data by area, it will at least permit us to identify tracts with the largest proportion of youngsters. Follow the steps below to display proportional concentrations of children age six and under.

1. Click on the *Tracts* theme in the legend area to make it active.

2. Pull down the Edit menu from the menu bar, and select the Copy Theme option. This procedure copies the theme onto the clipboard; select the Paste option from the Edit menu to create a second instance of this theme in the view.

3. An exact copy is added to the top of the Table of Contents. To differentiate the copy from the original, we will rename the copy. This is accomplished by clicking on the theme entry in the Table of Contents to make it the active theme, and then selecting Properties from the Theme menu. For Theme Name, change *Tracts* to *Tracts - by Percent*. Click OK to accept the change.

4. At this moment, the second theme is indistinguishable from the first with the exception of the theme names. To remedy the situation, double-click on the theme to bring up the Legend Editor. Classify the theme as you did the first time around, this time selecting *Percent_le6* from the Field pull-down menu. An initial classification of five classes appears, the highest of which is 11 - 13. For the current analysis, we want an either/or classification. Select Classify from the Legend Editor, and then enter *2* for number of classes in the classification menu.

Chapter 2: The Whirlwind Tour

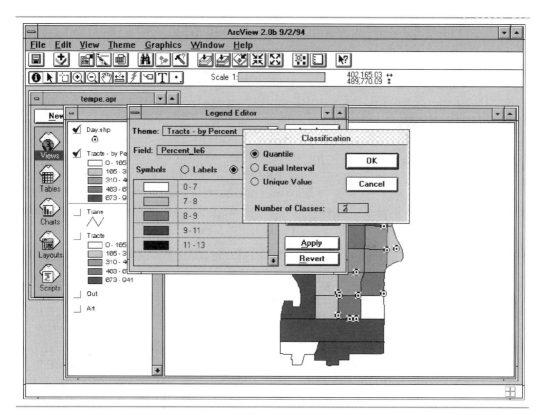

Preparing to generate two classes on Percent_le6.

5. In the Legend Editor, we see two classes: 0 - 9, and 9 - 13. For a threshold of ten percent, click on Values and enter the following new class ranges: 0 - 10, and 10 - 13. Click on Apply. The new classes are applied to the theme, but the Table of Contents still shows the original classes. To update the Table of Contents, click on Labels, and enter 0 - 10 and 10 - 13 for the two class labels. Click on Apply.

6. Change the symbology for the two classes. Double-click on the Symbol to bring up the Palette choices. For the first, choose the upper left symbol pattern, the open square. This is a transparent symbol and will allow other data to draw through it. For the second symbol, choose a bold diagonal hatch pattern which will be readily distinguishable against the solid shades of the population theme. When you click on Apply, the changes are applied to the theme.

Exercise 1: A Sample Application 45

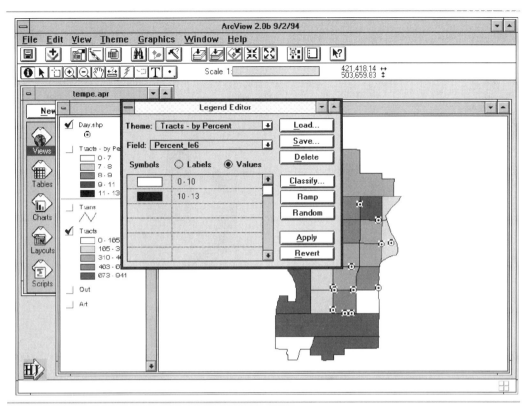

The revised classes for Percent_le6.

7. Click on (activate) the theme in the Table of Contents to display the new theme along with the previous themes. You will see the map appearing in the following illustration. All but one block with the highest raw counts for population age six years and under also have a high proportion (greater than 10%) of the population in this age group. Southwest Tempe may well be a favorable area for Sharon to locate her new day care center. (The incremental project has been saved as *ch2fin.apr*.)

46 Chapter 2: The Whirlwind Tour

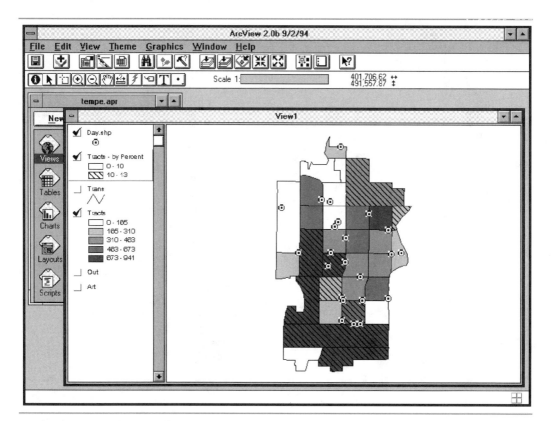

The final distribution of day care locations and population age six and under.

Before finalizing our proposal for Sharon's location options, we might want to explore additional variables, such as household income, or converting the population age six and under to a per acre figure. You have the data you need to extend the analysis. Later, when you read about how to add fields and edit records in a table, you can explore these additional analyses on your own.

Chapter 3
Getting Started: Projects and Views

In Chapter 1, you were introduced to ArcView and the fundamentals of GIS. In Chapter 2, we took you on a whirlwind tour. At this point, we suspect that either the light bulb has clicked on or we have you totally lost. We hope it's the former, but if the latter applies, take heart—the rest of the book proceeds at a slower pace and covers the material in greater detail.

Where to begin? When you are lost, you ask for help. With a pull-down Help menu and the HelpTool icon, help is always just around the corner in ArcView.

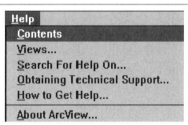

The pull-down Help menu.

The HelpTool icon.

Like many Windows-based products, ArcView's help system provides guidance and support at both novice and expert levels. The system presents ArcView's features and provides examples on usage. The HelpTool icon is used primarily to obtain information about a specific tool or menu option. Clicking on HelpTool followed by clicking on the tool in question brings forth a help page for that tool. Alternatively, you can click on a window or element to bring forth the help page.

The pull-down Help menu is particularly useful when you are not certain about which function to perform, how a function operates, the steps to take, or the options you have. The Help menu allows you to browse or search the help system via a keyword or phrase.

Remember that whenever you are inside ArcView's help system, you can print pages for later reference.

The First Step: Opening a Project

Efficient navigation in ArcView requires understanding of the program's organizational structure. The highest organizational unit in ArcView is the *project*. An ArcView project is similar to a working file in that it allows you to group all program components—*views, themes, tables, charts, layouts*, and *scripts*—into a single unit.

An ArcView project is stored as a project file. Basically, this file stores the ArcView environment exactly as it is at the time you execute Save, and enables you to recall the session at a later time. All ArcView components, including joined tables, logical or spatial reselects, geocoded event themes, as well as references to all the graphic files and maps on your screen, are saved in the project file. All components will be covered in subsequent chapters. The key point is that an ArcView project, stored as an ASCII file with a *.apr* extension, is the repository for your working environment.

Project files are *dynamic*: the steps that led to the views of your data are stored rather than the data itself. As a result, the project stays current with your data. The next time you open an ArcView project, any changes in your data will be reflected in your maps. Database operations, such as tabular joins or logical queries will be performed anew. In this way, ArcView ensures that what you see is always updated. The ArcView project is an excellent tool for tracking and modeling a dynamic database.

To create a new project from scratch, begin by selecting New Project from the File menu. The Open Project selection from the same menu is used to open an existing project.

The First Step: Opening a Project 49

✔ ***TIP:*** *Because the UNIX version of ArcView is started from the command line, you can specify the name of the project to open when starting ArcView (e.g., arcview myproj.apr).*

As demonstrated in Exercise 1, the first steps in a new ArcView project are opening a *view* (essentially a map), and adding data. To open a view, click on the Views icon from the project window and select the New option. A view is opened, with the default name of *View1*.

Once a view is opened, you can add data. In ArcView, data fall into two main categories: spatial and tabular.

Spatial data are also referred to as *digital cartographic data.* Think of this data as specialized graphics in which digitized geographic elements are coded with the absolute coordinates that locate the elements on Earth. Spatial data are depicted in vector and raster forms. ArcView supports both.

In *vector* data, map elements are stored as a series of x,y coordinates that define lines, points and polygons. ArcView supports vector data in two formats: ARC/INFO coverages and ArcView shape files. ARC/INFO data support is especially important because ARC/INFO can convert data from myriad other sources, which are therefore translatable to ArcView.

➥ ***NOTE:*** *The shape file is ArcView's native data format, and has been optimized for performance within the program. See "Working with Shape Files" in Chapter 9.*

In *raster* data, a geographic area is divided into rows and columns, effectively creating a data grid. Each grid cell, or *pixel,* is coded with a value representing information at the location being depicted. The most familiar raster data type is satellite imagery. Raster data are also referred to as *image* data. Raster formats supported in ArcView include the following:

- ARC/INFO Grid data
- TIFF
- ERDAS
- BSQ, BIL, and BIP
- Sun raster files
- Run-length compressed files

Once spatial data are accounted for, it is then desirable to bring in descriptive or tabular data that can be linked to your graphics. Typically, tabular data are imported in order to provide additional information about an existing spatial data set or *theme* (geography). Tabular data may be geographic in nature, having a locational component such as an address or zip code, or it may be informational, containing additional attributes associated with features in a theme, such as soil properties or land use descriptions. (See insert titled "Where to Get Data.") Standard tabular formats supported in ArcView include the following:

- dBase files
- SQL servers such as Oracle, Sybase, Ingres and Informix
- ASCII tab- or comma-delimited text files
- INFO tables

Where To Get Data

To a large extent, your data sources will depend on where you work and how many of your colleagues are already using ARC/INFO or other ESRI products. ArcView users will generally obtain data from the following five sources:

- Data supplied with the ArcView software. ArcView ships with the digital five-digit zip code boundaries for the entire United States.

- ARC/INFO users within your own organization. Not only will they be a likely source of existing data, they will also be able to provide powerful data conversion capabilities as well as the ability to capture data from scratch by digitizing or scanning.

- Public agencies, such as local utilities and federal, state, county or municipal governments. These agencies are traditional GIS strongholds. Many have been creating data for years. In addition, because of government policy on public access, this data is often available at nominal expense

❏ Colleges and universities. While universities do not typically provide a GIS production shop with an extensive data catalog, they can serve as a valuable source of expertise for GIS data capture and conversion for specific projects. Student interns are often available to assist, not only in initial data capture, but in the subsequent analysis as well.

❏ Private data vendors. These firms may be regional or national in scope, and range from basic providers of geographic data to full service consulting shops capable of handling all aspects of GIS project management.

While public agencies are a likely first place to look—particularly if you need government information such as census, planning and zoning, or natural resource data—you may find yourself on your own with regard to locating data sources and ensuring that the data is provided in a usable format. Private consultants and data resellers, by providing government as well as proprietary data on a value-added basis, can help ensure that data formatting, such as map coordinates and map projections, is consistent across all data sets, and that the associated tabular data is formatted appropriately for joining to your own tabular data.

ESRI provides data and consulting services, as does Equifax/National Decision Systems (NDS), the suppliers of the market and demographic data used in our examples. Many other private suppliers are also available, including Wessex, SMI, Claritas/National Planning Data Corporation, Etak, Geographic Data Technology (GDT), and Business Location Research (BLR). In addition, many national and regional consulting firms are now providing GIS data and services, thereby increasing the likelihood you can find a suitable firm close to your place of business. The *ArcData* catalog, which ships with ArcView, provides a comprehensive listing of spatial data available to support a wide range of ArcView applications.

Finally, do not overlook your local ARC/INFO users group. Many state and regional user groups are very active, and can assist you in locating data or providing short-term expertise. Contact your regional ESRI office for information on user groups active in your area.

Adding Data

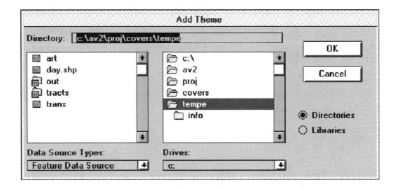

The Add Theme dialog window.

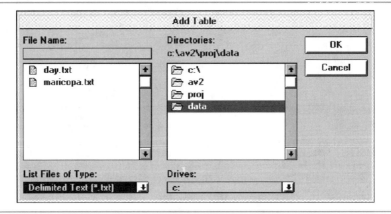

The Add Table dialog window.

Whether you are adding spatial data (themes) or tabular data (tables), the dialog window looks very similar. The functions, however, are accessed from ArcView quite differently.

The Add Theme option is available only when working with Views (maps). Both new and existing maps are opened via the Views icon in the project window. Once the view is open, new geographies (themes), can be added to the map via the Add Theme icon on the button bar or via the View menu from the menu bar. The corresponding dialog asks the user to identify spatial files through the standard Windows template requesting drives and directories, with two additional choices: a toggle between Directories and Libraries, and identification of the data source type.

Adding Data 53

The toggle defaults to Directories, but indicates that the user can also select to work with Libraries, a special ARC/INFO storage format. (For more information on Libraries, see the "Address Events" section in Chapter 4.) The data source selection allows you to view available vector data files and grid and image (raster) files. Vector data include ARC/INFO coverages and ArcView shape files. (See insert titled "Importing and Exporting ARC/INFO Data.")

Unlike graphic data, tabular data is accessed from the project window rather than the active *view*. To understand this, it is worthwhile to review the difference between how ArcView sees views and tables. Recall that views are equivalent to maps. Maps contain several layers of data, such as streets, political boundaries, lakes, and points of interest. Each layer becomes a theme in ArcView. All layers or themes combined are represented as a view.

Views are not exclusive. The same theme can be present in several different views. However, only one view can be active at a time.

There is no corresponding structure for *tables*. Tables are stored in a common area. There are no linkages representing how tables appear, nor how tables are combined or juxtaposed with other tables. Hence, adding tables is accomplished from the project window, rather than the view window.

Adding tabular data with the Project pull-down menu.

The Add Tables dialog window presents a pull-down menu that allows you to specify the file type you wish to open. File types follow:

- ❒ dBase (.dbf)
- ❒ Delimited text (.txt)
- ❒ INFO

54 Chapter 3: Getting Started: Projects and Views

If the user does not select a tabular data file type, ArcView will choose dBase as the default. To open a tabular data source, highlight the file name and click OK.

More About Themes

In ArcView, themes are the basic building blocks of the system. Themes can represent essentially any spatial data set, that is, features with locational attributes. Data sources that can be represented as themes include ARC/INFO coverages and images, as well as tabular data with locational components, such as latitude/longitude coordinates or street addresses.

Note that ArcView themes need not represent all features from the original data source. For example, an ARC/INFO coverage may be comprised of both line features and polygon features, such as a census coverage containing both street attributes and census tract attributes. An ArcView theme could represent either lines or polygonal elements from the ARC/INFO coverage. When working with ARC/INFO data containing more than one class of features, the features will be represented with a folder icon in the Add Theme dialog window. To access one of the coverage's feature types, click on the folder icon to display the feature choices within.

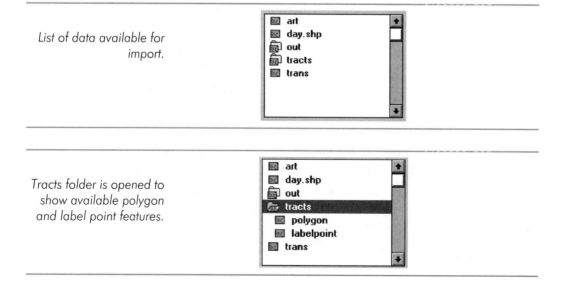

List of data available for import.

Tracts folder is opened to show available polygon and label point features.

A theme might also be created from a subset of feature elements in the original data source. By using query techniques, you can create a theme representing only a subset of total features in a data source. For example, parks or recreational areas could be extracted from an ARC/INFO land use coverage. We will explore this topic further in Chapter 6.

In working with themes, you can set properties that describe how the themes are named, how they appear and other special processing rules. To access the dialog window for setting theme properties, click on the name of the theme in the Table of Contents to make it active. Next, click on the Theme Properties icon from the button bar or select Properties from the Theme menu. Theme properties are listed below. Setting theme properties is included in the exercise at the end of this chapter.

- Theme name
- Selected theme features
- Minimum and maximum scales at which the theme will be displayed
- Field and position for labeling theme elements
- Theme hot links to other data sources or applications
- Theme geocoding properties

More About Views

Views organize themes. If themes are the individual players on your ArcView team, views are the stadiums in which they perform.

Each view window contains a graphics display area and a Table of Contents which lists all themes present in the view. The entry for each theme in the Table of Contents contains the name of the theme, the symbol legend for the theme, and a check box indicating whether the theme is currently set to draw in the view. A scroll bar is present in case the list of themes is longer than the view window.

The drawing order for themes in a view is determined by the position of the themes in the Table of Contents. Themes at the bottom are drawn first, and themes at the top are drawn last. Thus, if you want a particular theme to be "on top" of another, position it accordingly in the Table of

Contents. To change the theme order, click on and drag the entry for the theme to a new position among the themes listed in the Table of Contents.

Many of ArcView's operations, such as selecting and identifying features or zooming in and out, are configured to work on an active theme. To make a theme active, click on its name in the Table of Contents. The theme will appear to be highlighted, or raised, in the legend.

Similar to themes, the properties of views can also be set. To set view properties, select Properties from the View menu. The dialog box contains the following properties:

- Name of the view
- Map units
- Display units
- Map projection

Saving Your Work

There are three ways to save an existing project as you work: select Save from the File menu; click on the Save Project icon from the button bar; or key in <Ctrl>+S.

ArcView projects can be saved at any time. Upon issuing a Save, environment components (themes, tables, charts, layouts, and scripts), and all dynamic aspects of the components (joined tables, logical queries, and thematic display) are written to the project file. These will be restored in exactly the same state when you reopen the project.

Note that you can also create a new project from an existing one by saving the project to a different name. Click the project window title bar, choose Save Project As from the pull-down File menu, and enter the new project name.

The concepts in this chapter are critical to establishing a strong understanding of ArcView. Some of these concepts are reinforced in Exercise 2.

Exercise 2: Opening a Project

Because this is a new project, we will be starting from scratch. Start ArcView, and from the opening File menu, select New Project. With the project opened, click on the Views icon to make it active, and select New to open a new view.

A view window is opened, with the title *View1*. Resize the application and view windows as desired to accommodate the graphics displayed during the exercise.

Set the working directory to the current workspace. Click on the project window to make it active, and select Properties from the Project menu. For *Work Directory*, enter *$IAPATH\work*. Click OK to accept the change, and click on the *View1* window to make it active.

The current project involves locating potential markets for a client. To accomplish this, we have obtained three data sets: demographic, cable viewership, and restaurant locations.

The first task is to load the spatial data (geography) we will be using to locate and reference the tabular data. For this project, we will use two spatial themes: census block groups, and the census street net. The census block groups will be used to display demographic and cable viewer data, which has been aggregated to the block group level. The street net will be used as reference for displaying restaurant data, which has been coded with x,y coordinate locations.

Importing and Exporting ARC/INFO Data

The initial step in setting up an ArcView project is loading data, such as ARC/INFO coverages. Typically, these coverages were created in ARC/INFO on a UNIX-based workstation. Methods for transferring a coverage composed on a UNIX platform to a PC depend on the data source, and, to a lesser degree, the maintenance needs for your data. To explain the options properly, we need to get technical.

An ARC/INFO coverage consists of data files containing the spatial data and the associated attribute data. In the workstation version of ARC/INFO, the attribute data is stored in the INFO database. For each ARC/INFO coverage, a directory holds the data files. In addition, an INFO directory holds the data file templates for the attribute data. Both directories comprise the ARC/INFO *workspace*.

ArcView can directly access an ARC/INFO workspace. Access can occur through a network of workstations and PCs, or the workspace can be copied as a unit to the PC. File names, however, must adhere to the DOS 8.3 standard. (The standard dictates that the file name contains eight characters or less, and the extension, three characters or less. An example is *filename.dbf*.)

As of ARC/INFO 7.0, all new workspaces are created to conform to the DOS 8.3 standard. In addition, ARC/INFO's CONVERTWORKSPACE command converts a pre-7.0 workspace to the 7.0 format. Note, however, that the standard imposed in version 7.0 ensures that only the internal file names adhere to the DOS 8.3 standard. It is the user's responsibility to ensure that all ARC/INFO coverage names are eight characters or less in length.

An ARC/INFO workspace can be copied without prior conversion from a UNIX workstation to a DOS-based PC. (Note: Associated ASCII text files may need to be passed through a UNIX to DOS utility.) However, once the files are installed on the PC, there are no utilities within ArcView for maintenance of an ARC/INFO workspace. Due to the existence of a common INFO directory containing attribute table templates for all covers in the workspace, if an ARC/INFO cover needs to be updated, the entire workspace will need to be updated and reinstalled as a unit.

An alternative means of data exchange is via the IMPORT utility supplied with ArcView. The IMPORT utility reads an ARC/INFO EXPORT file—the ARC/INFO exchange format—and converts it to a PC ARC/INFO format workspace. In PC ARC/INFO, data files for each ARC/INFO cover are stored in a separate directory. Coverage attribute tables are stored in dBASE format. Along with table templates, the coverage attribute tables are stored in the coverage directory. Because there is no common INFO directory, each ARC/INFO coverage is stored in a single self-contained directory. Consequently, it is possible to update individual coverages in a workspace without having to recreate the entire workspace.

Certain limitations are inherent in ArcView's IMPORT utility under Windows. IMPORT does not support double-precision ARC/INFO coverages. Next, extended data types in ARC/INFO 7.0 (routes and regions) are not supported, nor are ARC/ INFO annotation TAT files. ARC/INFO covers with polygons containing more than 5000 vertices, while acceptable in ARC/INFO 7.0, will cause the ArcView IMPORT utility to abort. In addition, certain INFO item types are not supported, and any INFO REDEFINED items will be dropped.

> *NOTE: The IMPORT utility supplied with the UNIX version of ArcView is the same IMPORT utility supported within UNIX ARC/INFO. Accordingly, all ARC/INFO coverage feature classes and INFO items are supported.*

The ArcView shape file format can serve as an alternative exchange format when maintaining ARC/INFO workspaces on both platforms is not necessary. Shape files offer the advantage of uniform bi-directional translation between platforms without the limitations mentioned above. However, certain limitations still remain. Shape files can be created only from a single feature class of an ARC/INFO coverage; coverages containing multiple feature classes, such as arc and polygon attributes, must be translated into separate shape files. In addition, associated cover attribute tables beyond the feature attribute table, such as look-up tables, must be exported and transferred separately. Next, if you wish to maintain coincident workspaces across platforms so that a single ArcView project file can be used on both platforms, the data must be maintained on the ARC/INFO platform in the shape file format. This procedure necessitates redundant data storage.

Which method is best for you? The choice depends on your data requirements, the need to maintain ARC/INFO coverages, and the ability to maintain a mirror of your PC ARC/INFO workspace on a UNIX workstation. While maintaining your ARC/INFO workspace in UNIX format makes maintenance and updating more difficult, it does allow you the full range of ARC/INFO data types. In addition, workspaces can be copied back to the workstation as needed without conversion.

The PC ARC/INFO format workspace created by the IMPORT utility cannot easily be transferred back to the workstation for maintenance. Workstation UNIX-based ARC/INFO cannot read a PC ARC/INFO workspace, and there is no EXPORT utility in ArcView to create an ARC/INFO EXPORT file from a coverage modified in ArcView. To transfer a PC ARC/INFO workspace to a UNIX based ARC/INFO installation, it is necessary to copy the workspace to a PC where PC ARC/INFO is installed, and use the PC ARC/INFO EXPORT utility to create an EXPORT file which can subsequently be read by UNIX-based ARC/INFO.

> NOTE: As mentioned previously, the ArcView shape file format offers ease of data exchange at the expense of maintaining redundant data storage. In addition, dealing with ARC/INFO coverages containing multiple feature classes may complicate data exchange.

If you obtain your data from a third party provider—and the data has been formatted for use in ArcView, and the vendor has already provided for maintenance of your data sets—, none of the above necessarily applies.

Steps for loading spatial and tabular data appear below.

1. Click on the Add Themes icon from the button bar.

2. Navigate to the *$IAPATH\data* directory, and select the *blkgrp* and *trans* shape files. Display the resulting themes by clicking on the box next to the theme name in the View Table of Contents.

Feel free to call up the legend editor by double-clicking on the theme name in the Table of Contents if you wish to change the theme's color or symbology.

3. Once you are satisfied that the spatial data has been imported successfully, add the tabular data. Click on the project window (*Untitled*), and select Add Table from the Project pull-down menu. Navigate to the *$IAPATH\data* directory, and set the file type to

Exercise 2: Opening a Project 61

display dBASE (*.dbf). From this list, select *demog.dbf, cable.dbf* and *restrnt.dbf.* Remember, after selecting the first file with the mouse, you can add to your selected set by holding down the <Shift> key while clicking on your selection. Click on OK to add these tables to the project. As the tables are added, they are opened for display, each in a separate window. (The incremental project has been saved as *ch3a.apr.*)

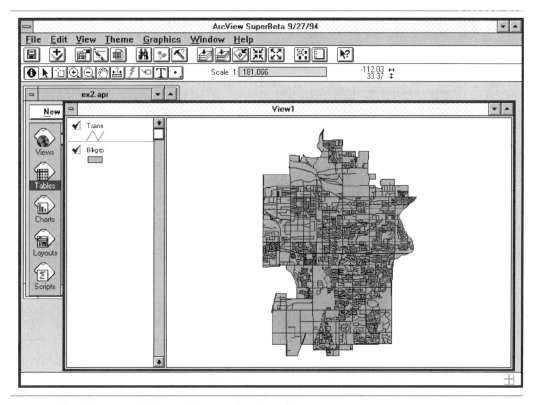

The Trans and Blkgrp themes displayed.

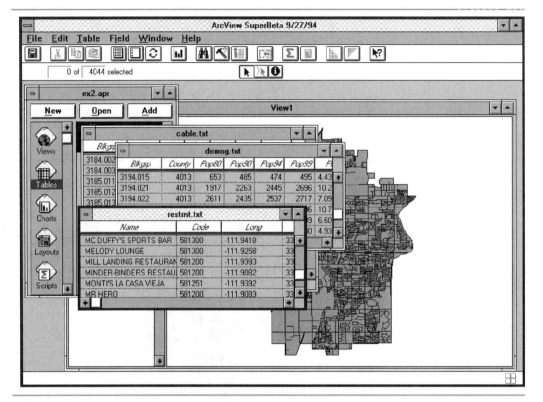

The cable.dbf, demog.dbf and restrnt.dbf tables displayed.

The linkages between the tabular data and the spatial themes—the starting point for our geographic analysis—will be established in Exercise 3 (Chapter 4). Before closing the project, let's take a moment to examine certain properties of view and theme displays.

Theme Display

As mentioned earlier, the drawing order for themes is from the bottom to the top of the theme listing in the Table of Contents. To change the drawing order, click on the theme entry in the Table of Contents, and while holding down the mouse button, drag it to a new position.

Change the drawing order of the themes in the exercise by first clicking on the check boxes so that both the *trans* and *blkgrp* themes are displayed. Click and drag the *blkgrp* theme so that it is at the top of the list. Note that the solid shading for the *blkgrp* polygons draws over the street net from the *Trans* theme.

There are two ways to allow the *blkgrp* theme to be displayed on top of the *trans* theme while still allowing the *trans* theme to be seen. One method involves changing the shade symbol from a solid to a hatched pattern, and the other is to change the theme symbology so that only the polygon outlines (block group boundaries) are drawn. The symbology for a theme is changed by accessing the Legend Editor.

To access the Legend Editor, double-click on the theme entry in the Table of Contents. This action brings forth the Legend Editor window. To change the symbol pattern and color, double-click on the symbol box in the legend to access the Palette Editor. Note that the current shade pattern selected is the solid black pattern which produces a solid fill of the selected color. Click on one of the hatched patterns, and apply this change by clicking on the Apply button in the Legend Editor window. The theme is immediately redrawn with the new symbol and the *Trans* theme can be seen.

To change the symbol color, click on the brush icon in the palette window to switch to the Color Palette. Click on the desired color, and Apply as before.

To draw theme outlines only, click on the far left icon in the Palette window to return to the Palette Editor. Select the clear symbol at the upper far left.

This symbol pattern is transparent, which allows the polygons from the theme to be drawn without shading. At the bottom of the Palette Editor from the pull-down list for Outline, select 2. The line weight is changed to a double width.

64   Chapter 3: Getting Started: Projects and Views

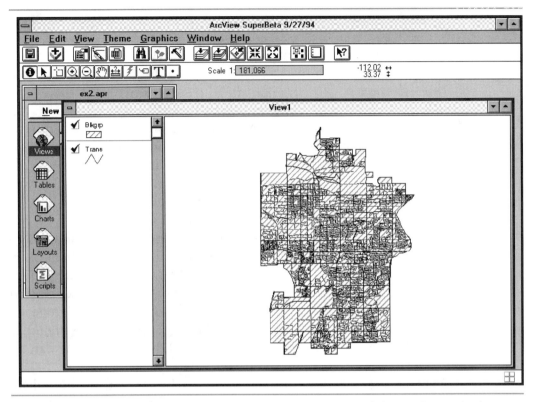

Diagonal hatching allowing the Blkgrp theme to be displayed over the Trans theme.

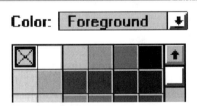

The clear symbol displayed in the upper left corner of the Color Palette window.

The polygons are now drawn with a bold outline. To change the color, return to the Color Palette, and from the pull-down list for Color (which defaults to Foreground), select Outline.

Selected color choices will now apply only to the polygon borders.

Exercise 2: Opening a Project

Selecting 2 for a double width line from the Outline pull-down menu in the Palette Editor window.

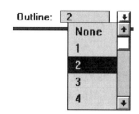

Selecting Outline from the Color pull-down menu in the Color Palette window.

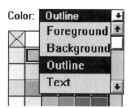

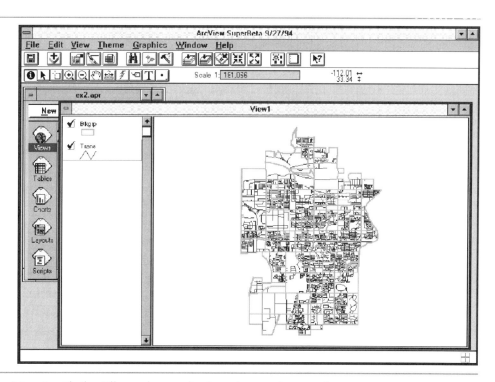

View1 with the Blkgrp theme displayed as outlines only.

View Display

When a view is initially opened, the map units are not set. In order for all themes to be displayed together, the themes you add must share the same coordinate system as the view. Once you have added themes to the view, you should then set the view's map units.

From the View pull-down on the menu bar, select Properties. A scrolling list appears for Map Units, which defaults to Unknown. (Map units are the units in which the coordinates of your spatial data are stored.) Click on the downward arrow to activate the list.

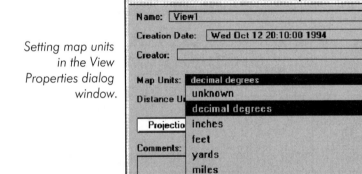

Setting map units in the View Properties dialog window.

The themes we have added are stored using latitude and longitude coordinates. Latitude and longitude are referenced in degrees, minutes, and seconds, and are commonly calculated as decimal degrees. Select *decimal degrees* from the list, and click OK.

Note that a numerical value is now displayed in the Scale box on the tool bar. Click on the Zoom Out tool from the tool bar, and click in the display area to zoom out from a selected point. As the map is redrawn, note how the scale value increases. Now click on the Zoom to All Themes icon from the button bar. The display is redrawn at the full extent of the theme, at a scale of approximately 1:175,000. 1: 140,619

Knowledge of the map scale can be important in two areas. First, the Scale box allows you to control the display by direct input. Click on the Scale box, and change the scale number to read 50000. Upon hitting

Exercise 2: Opening a Project 67

<Enter>, the map is redrawn at the new scale—either zooming in or out from the center of your previous display.

Second, the scale can be used to control the display of a specific theme. For example, click on the *Trans* theme to make it active, and select Properties from the Theme pull-down menu to access the Theme Properties dialog window. Click on the Display icon on the left side of the window. You are presented with two lines on which to enter minimum and maximum scale values.

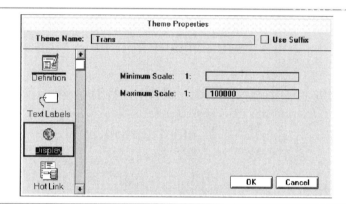

Setting the maximum display scale in the Theme Properties dialog window.

Key in 100000 for the maximum scale, and click on OK. The display is redrawn. Zoom out to the full extent of the theme by clicking on the Zoom to All Themes icon from the button bar. Note that even though the theme is clicked on in the Table of Contents, the theme is not drawn. The value displayed in the scale bar shows approximately 175000, greater than the 100000 value we entered for the maximum display scale. Therefore, this theme is not displayed. In this manner you can create scale-dependent themes. Thematic display can accordingly be limited when the display would not serve a useful purpose, such as when zoomed to a map extent where the amount of detail renders the display illegible, or redraw time extensive.

We have now explored the basics on adding themes and tables to a view, as well as the basics on controlling the properties with which themes are displayed. Remember to save your work before exiting because in the next exercise we will be continuing where we finished in this exercise. (The incremental project has been saved as *ch3fin.apr.*)

Chapter 4

Extending Data

In Chapter 3, we focused on the basics of themes and views, and importing your data into ArcView. Before we can go on to maps, queries and analysis, there is still more setup work to do. As demonstrated in Exercise 1 (Chapter 2), linking tabular data to spatial themes and creating a geocoded event theme were necessary to make the data usable. In this chapter we will cover additional basics in extending data.

We must first take a peek under the hood to deal with data formatting of map projections. ArcView requires all spatial data sources to share a common coordinate system. If you have obtained all of your data from a single source, such as a third party data provider, or if you are using ArcView with an existing ARC/INFO installation, this issue has most likely been taken care of for you. However, if you are gathering data on your own from several different sources, we encourage you to study the insert titled "Defining a Common Ground: Dealing with Map Projections."

Defining a Common Ground: Dealing with Map Projections

In Chapter 3 we mentioned that in digital cartographic data, map elements are stored as a series of x,y coordinates which represent location on Earth's surface. In Exercise 2, we brought in spatial data, and set map units to latitude/longitude expressed as decimal degrees. These activities, however, merit some discussion. While we can locate a point on Earth with great accuracy, representing the same point on a map is still an approximation.

In attempting to use a flat surface (map) to represent a curved surface (Earth), distortion is inevitable.

Geographers have developed a number of *map projections* to reduce distortion. While a full discussion of map projections is beyond the scope of this text, suffice it to say that each projection strives to resolve the distortion problem in a manner that satisfies the needs of a particular subset of users. Some projections are very accurate over short distances; others are better for long distances, or representing large areas of the planet. Some projections are optimal for representing distances on a map, and others are optimized to accurately represent area.

The coordinate system used in this book's exercise sample data is latitude/longitude, often referred to as *geographic coordinates*. Strictly speaking, this system is not a map projection but rather a reference grid. Distances along the grid are uniform on the y axis as we travel toward the poles. In contrast, distances along the grid on the x axis become increasingly shorter as we progress toward the poles. Because the grid is uniform, it is relatively easy to mathematically transform geographic coordinates to another map projection. As such, latitude/longitude coordinates are useful as a common reference system for storing digital spatial data when the ultimate use of the data is not known in advance. All GIS and desktop mapping software, including ArcView and ARC/INFO, can utilize data stored in latitude/longitude form.

The two key points we would like to make about map projections appear below.

❑ Not all available digital data is stored in the same map coordinates and projections.

❑ ArcView requires that all digital data share the same map coordinates and projections.

In the United States, in addition to latitude/longitude, digital data is likely to be found in the Universal Transverse Mercator (UTM) or State Plane Coordinates map projections. Both projections divide the United States into a series of zones, each with its own relative x and y coordinates. For example, the state of Arizona is covered by UTM zones 11 and 12, and the state planezones of Arizona West, Arizona Central and Arizona East. Both projections are very accurate in the mapping of regional extent. The UTM projection is widely used by public agencies, particularly those involved in naturalresource management such as the U.S. Geological Survey and the Bureau of Land Management. The State Plane Coordinate system is commonly used by land surveyors, public works departments, and state and local transportation departments.

ESRI's ARC/INFO software can transform data to and from all major map projections. ArcView cannot. If possible, request that your data be projected into a common coordinate system before delivery in order to avoid seeking a data conversion source at a later date.

In addition to map projections which fit the Earth's surface to a flat map, we also need to compensate for the fact that the Earth is not a true sphere. The establishment of a network of precisely located points with respect to both location and elevation (i.e., horizontal and vertical) is referred to as *geodetic control*. With geodetic control, the accurate location of points on the Earth's surface is accomplished with the aid of a *datum* which accurately describes the shape of the Earth. In the United States, the two datums in general use are the North American Datum of 1927, and the North American Datum of 1983. The vast majority of federal and regional mapping projects, and consequently, the majority of digital cartographic data derived from these projects, were carried out using the 1927 datum. At present, however, the 1927 datum is being adjusted to the 1983 datum. (Between the 1927 and 1983 datums, the shape of the Earth is redefined by adjusting all coordinates which define the shape.) In turn, corresponding digital databases will need to be adjusted as well. This is particularly significant in mapping projects utilizing global positioning system (satellite) survey units. All points located using GPS are tied to the 1983 datum.

Lest you think that the difference between the 1927 and 1983 geodetic control devices is insignificant to all but surveyors, the distance between points from the 1927 to the 1983 datum can be in excess of 90 feet. This difference can certainly be significant when locating features such as water mains or property lines. If your mapping project requires you to maintain high levels of accuracy, you will need to ensure that the datum is specified as well as when the digital data was obtained.

Joining Tables

The ability to join tables based on a common item is one of the most important functions in database management. Simply stated, it allows for non-redundant data storage and simplifies database maintenance.

For example, a data file containing the inventory of a large parts warehouse can include a price code field for each item. A look-up table can then be prepared associating each price code with the current price. Editing the much smaller look-up table ensures that price changes will be subsequently applied to each item referencing particular price codes. The two major benefits deriving from the use of joined tables are (1) when changes occur, only one file has to be updated; and (2) because the link between joined tables is dynamic, subsequent views of the joined table will reflect changes after a file has been changed.

The strengths of the above model also apply to GIS database design. For instance, a theme of digitized land parcel boundaries can be prepared and coded with the assessor's parcel number. The theme can be joined via the same field to tabular data from the assessor's office. In this manner frequent changes in parcel attributes, such as assessed value or last sale, can be kept separate from the digitized parcel boundaries, which change much less frequently. The join ensures that changes to the joined table will automatically be associated with the spatial theme, keeping mapped attributes current and also allowing for temporal change mapping.

ArcView's Join function is particularly robust, allowing for tables from dissimilar sources to be joined and stored as a *virtual table*. The source file for joining to the spatial theme attribute table can be a dBase or INFO table, a table from an RDBMS (relational database management system) such as Oracle or Sybase, or a delimited ASCII text file. Once these tables are imported into ArcView, they are stored in the same internal format and are available for further manipulation, including joining to spatial theme attribute tables.

The mechanics of how to perform a join are straightforward, and will be covered in Exercise 3 at the end of this chapter. A related concept, the relationship between the source table and the destination table, remains to be covered.

In ArcView, the *destination table* is the table to which the fields from the *source table* will be appended. The destination table is typically the attribute table for the spatial theme. The results of a join are accurate only if there is a one-to-one correspondence between records being matched from the source to the destination table; that is, only one unique record in the source exists for each record in the destination. The relationship may be one-to-one, as in the link between a land parcel theme and the associated parcel data. The relationship may also be a many-to-one relationship, as in the link between a land use theme and the look-up table explaining each land use code.

If there are many records in the source table linking to the destination table, only the first record from the source table will be joined to the destination table. In this situation, you should be *linking* rather than joining the tables.

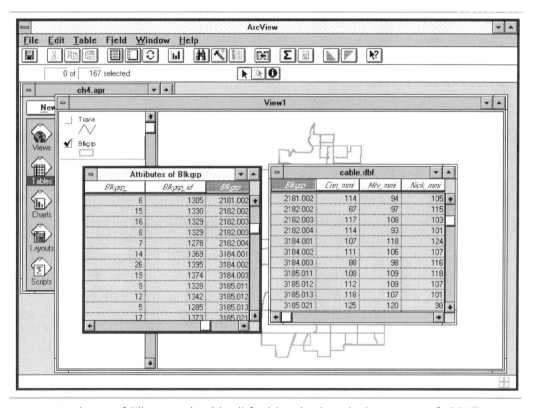

Attributes of Blkgrp and cable.dbf tables displayed, showing join field Blkgrp with
  common values.

Join Versus Link

The Link function is used when a many-to-one relationship exists between the source table and the destination table. Linking the two tables ensures that all records from the source table are associated with the destination table. The source table records will subsequently be available when querying the spatial theme. For example, clicking on a land parcel would bring up all records from a linked table comprised of past owners of that parcel.

> **NOTE**: *Unlike the Join function, Link merely establishes a link between tables rather than joining the tables into a new virtual table. As such, fields in the linked table are not available for thematic query or analysis. Unless a many-to-one relationship exists between the source table and the destination table, you should always use Join.*

Event Themes

In Exercise 1 (Chapter 2), you may recall how we located day care centers against a street net theme geocoded by street address. The resulting point theme was one example of an ArcView *event theme*.

In ArcView, an event theme is constructed from an *event table* which contains geographic locations. These locations can be absolute, such as x,y coordinates, or relative, such as street addresses.

When an event theme is created, the geographic locations from the event table are converted into an ArcView-supported spatial data format. In a table containing absolute locations, ArcView associates a point on the theme with every x,y coordinate pair. In a table containing relative locations, ArcView creates a *shape file*—ArcView's native spatial data format—containing point or linear features corresponding to the location of each entry in the event table. In brief, ArcView translates each feature from a relative location to an absolute location, and makes these features available for subsequent query and analysis.

ArcView supports the following event categories:

- ❏ XY events
- ❏ Route events
- ❏ Address events

XY Events

XY event tables contain the exact location of point features using x,y coordinates. The map coordinate system should correspond to the spatial themes against which these events are to be displayed. Commonly used systems include latitude/longitude, UTM, and State Plane Coordinates.

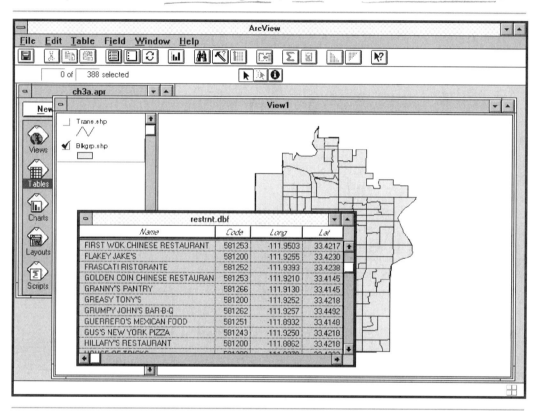

Restrnt.dbf displayed showing geocoded fields Long and Lat.

> ❧ **NOTE**: *If you are entering latitude/longitude coordinates, verify that the coordinates are properly coded as negative or positive to match the geographic quadrant in which they are located. For example, the longitude values for all locations in the northwest quadrant, which includes the United States, should be entered as negative values. Typically, these values are left unsigned when data is gathered; thus a longitude-latitude (x,y) coordinate pair entered as 112.30 33.12 should properly have been coded as -112.30 33.12.*

Importing these coordinates unsigned would result in a rather strange distortion—the resulting theme will appear to be the inverse or mirror image of what it was intended to portray.

Route Events

Route event tables contain the relative location of features along a *route system*. Route systems are most commonly associated with road networks, and are referenced as a distance from a known starting point, such as 12.3 miles from the beginning of Route 5.

> **NOTE:** *A route system must be built in ARC/INFO before route event themes can be added in ArcView.*

Route events can take the following forms:

- ***Point events*** are features located at specific points along a route, such as accident locations.

- ***Linear events*** are features located along a specific segment of a route. A pavement test section occurring from milepost 12.5 to milepost 13.8 along Route 5 is an example of a linear event.

- ***Continuous events*** are features located continuously along a route. An inventory of pavement condition by condition class rated as good, fair, or poor, and coded as to the location where the condition class changes is an example of continuous event data.

Address Events

Address event tables contain a locational identifier. The most common locational identifier is a street address along a linear street network. Address events can also be created against a polygon or point theme, such as a table containing Zip+4 data.

Exercise 1 included the basics of creating an event theme by matching street addresses against a geocoded street network. Regardless of the format of your particular address field, the steps followed in geocoding are the same. Geocoding steps follow:

Event Themes

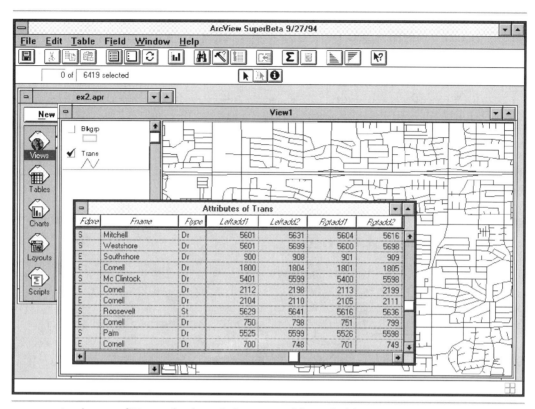

Attributes of Trans displayed showing address fields.

1. Add the required theme and table to the view. The spatial theme can be an ARC/INFO coverage or an ArcView shape file containing the requisite address fields for geocoding. The event table is comprised of fields containing address information.

2. Make the theme *matchable*. First, you need to make the theme active and set the theme's geocoding properties. Through the Geocoding Properties dialog window (accessed by choosing the Geocoding option from the Theme Properties window), select the address style and identify the fields containing the theme's address components.

3. Add the event theme. From the Add Event Theme dialog window accessed from the View menu, specify the name of the event table and the field in the table containing the address of the event. This action will call up the Geocoding Editor dialog window, which will

allow you to process the records in the event table singly or in batch, and edit records as necessary in order to ensure a match.

4. Add the geocoded theme to the view. When the matching is completed, an ArcView shape file is created containing point locations for each event in the event table. A theme based on this shape file can then be added to the view.

As you will discover when adding a geocoded theme, working with address events can be rather tricky. However, given the ubiquitous nature of address data, this is a skill worth mastering. For additional information on fine tuning address data, see the insert titled "The Science of Geocoding."

The Science of Geocoding

Street addresses are the most common form of geographic data. Nearly all of us work with addresses every day. Address geocoding in ArcView allows you to create a point theme based on address locations, which allows for each address point to be mapped. A 100 percent match of street addresses against a geocoded theme occurs only when both the address table and the geocoded street net are 100 percent properly coded. Perfect coding, alas, is quite rare.

There are ways, however, to improve your chances of obtaining a match between addresses and a geocoded theme.

Consider the following address in Phoenix: 1400 N. 16th Ave. Similar to many cities, Phoenix is laid out on a grid. Numbered streets run north and south, Avenues and Drives are located on the west side, and Streets and Places are located on the east side. Accordingly, it is vital that both the locational prefix and the categorical suffix be identified correctly in order to locate an address along a numbered street. Applying these rules, 1400 N. 16th Ave. falls in the northwest quadrant.

Let's examine the results when we locate this address against a geocoded street theme in ArcView. Keying in *1400 N 16th Ave* in the Locate dialog box causes the following matches to be displayed. (We have instructed ArcView to display all matches with a minimum score of 30. The score is given, followed by the record matched.)

99 - N 16th Ave
70 - S 16th Ave
67 - N 16th Dr
67 - N 16th St
38 - S 16th St

With all address fields present, the address will be properly located. Note, however, that if the 1400 block of N. 16th Ave. were not found in the street theme, the corresponding block on S. 16th Ave.—located many miles away in the southwest quadrant—would be preferentially matched over the corresponding block on N. 16th Dr., which is located only one block to the west.

If the street type field is missing, matching the remaining address, 1400 N 16th, produces the following matches:

83 - N 16th Dr
83 - N 16th Ave
83 - N 16th St
54 - S 16th Ave
54 - S 16th St

Note that as the score decreases for the best candidate (N. 16th Ave., from 99 to 83), the scores for the secondary candidates increase (N. 16th Dr. and N. 16th St., from 67 to 83). Leaving out just one component of your address can dramatically affect your matching success.

The completeness of the address can also affect the match. If we key in *1400 N 16th Av*, we obtain the same matches as if we had entered *Ave*. If, however, we key in *1400 N 16th Ae*, we obtain no matches. The lesson is clear: take care to see that your address data is coded accurately, and that the address prefixes and suffixes are coded in a consistent fashion with the format used in the street network.

Note that even painstaking quality control on address entry will not give you a match if the corresponding street in the street net is missing or incorrectly coded. In order to ensure a high percentage of matches, equal attention must be taken to guarantee that your geocoded street net is accurate and up-to-date.

> As mentioned in Chapter 2 (see insert, "Cleaning TIGER Street Nets"), the raw TIGER street files from the U.S. Census Bureau are likely to contain errors and omissions. The errors are not sufficient to render TIGER files unusable, but certainly enough to be aggravating. The Hutchinson-Daniel Law states that the street address missing from your theme is the street address you most need to match. It is possible, however, to edit the TIGER files to improve accuracy. The attribute table for the street net can be displayed and edited in ArcView. (See Chapter 9 for a discussion of table editing.) If you have a few addresses which do not match and an urgent deadline to meet, editing the attribute table can be enough to get you over the hump. Wholesale address changes, or digitizing new or revised streets may be beyond your resources. In this case, you might enlist the services of a consultant to clean up your street net, or purchase a revised street net from a commercial data provider.

One last note is in order before we proceed to the exercise on importing data into a project. The same strengths which make ArcView desirable as a stand-alone PC-based application make it attractive for use on a computer network in conjunction with ARC/INFO. There are, however, additional considerations to take into account when using ArcView in conjunction with ARC/INFO, particularly when accessing the network via ArcView running on a PC.

The primary concern involves file and directory naming conventions. In order to be reliably accessed from the PC, all directory and file names should adhere to the DOS 8.3 file naming convention (i.e., eight characters or less for the file name, and three or less for the file type extension, such as in *parkways.dbf*). Use this convention for naming ARC/INFO coverages on the network. Cover names in excess of eight characters may be truncated by the network protocol software with unpredictable results.

The internal file names for ARC/INFO coverages and INFO data files should also adhere to the DOS 8.3 naming convention. As of ARC/INFO 7.0, this naming convention is the default. There is also a command to convert any existing ARC/INFO workspaces to the 7.0 format. Note, however, that the 7.0 workspace format applies only to the naming of ARC/INFO and INFO internal files. It is up to the user to ensure that ARC/INFO coverage names do not exceed eight characters in length.

A noteworthy data source available to ArcView users networked to an ARC/INFO installation is *ARC/INFO Libraries*. At its most basic, this source is merely a formal directory structure allowing ARC/INFO coverages to be divided into tiles, each covering a portion of the total geographic area. A master index and master data templates allow for access of data across tile boundaries and ensure that the internal data format is kept consistent across tiles. As of version 7.0, ARC/INFO has extended the Library data model to a feature-based model, known as *ArcStorm Libraries*. ArcView accesses standard ARC/INFO Libraries and ArcStorm Libraries in the same manner.

Finally, note that the PC user accessing ARC/INFO libraries across a network will need to set an additional environment variable named *ARCHOME*, which contains the path to the ARC/INFO install directory. See the *Accessing ARC/INFO Libraries* entry in the ArcView help system for specifics on setting the environment variable.

Exercise 3: Extending Project Data

In Exercise 2, we imported the raw data (spatial themes and attribute tables) planned for use in subsequent exercises. In this exercise we will establish the linkages and create the event themes to make the data more usable, and begin to import primary research data.

Begin by opening the project saved in Chapter 3. If you did not complete Exercise 2, open the *ch4.apr* project file in the *$IAPATH\projects* directory. The following steps will establish a link between the demographic data and block groups.

1. Click on *blkgrp* in the view Table of Contents to make it the active theme.
2. Pull down the Edit menu and select Copy Themes. This action places a copy of the *blkgrp* theme on the clipboard.
3. Pull down the Edit menu again and select Paste. A copy of the theme will appear at the head of the view Table of Contents.

    ✒ *NOTE: The two-step copy/paste technique is similar to other Microsoft Windows applications, and also allows you to copy between views.*

82   Chapter 4: Extending Data

4. Click on the copied theme to make it the sole active theme. Select Properties from the Theme pull-down menu, and click on the Definition icon in the Theme Properties dialog window.

5. In the Theme Name box, remove the file name *blkgrp* and key in the new name, *demographics*. Click OK to accept the entry and close the Theme Properties dialog menu.

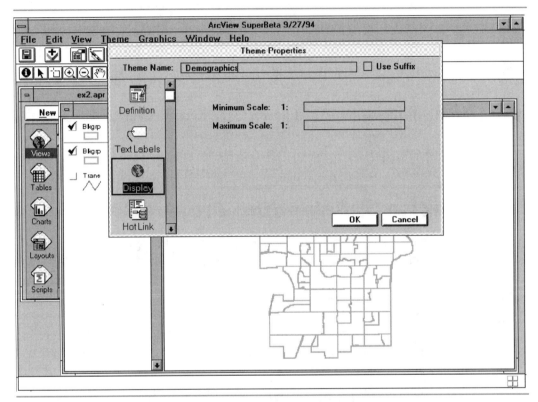

Renaming the copied Blkgrp theme to Demographics.

We are now ready to link the demographic data to the *Demographics* theme. This will be accomplished by joining the demographic table to the *Demographics* theme attribute table.

1. With the *Demographics* theme active, click on the Open Tables of the Active Themes icon from the button bar. The *Attributes of Demographics* table is opened in a window.

2. Click on the Views icon, and open the *demog.dbf* table.
   TABLES

Exercise 3: Extending Project Data 83

3. With both tables now visible, we need to select the common field that will be used to join the tables, that is, a field containing the same attributes in the same format in both tables. In the example, this field is the block group. In the *demog.dbf* table, the field is labeled *Blkgrp*, and in the *Attributes of Demographics* table, it is called *Bgc*.

4. Click on the *Blkgrp* field name below the title bar for the *demog.dbf* table. Note that the field is highlighted.

5. Click on the *Bgc* field name below the title bar for the *Attributes of Demographics* table. This field name is highlighted as well.

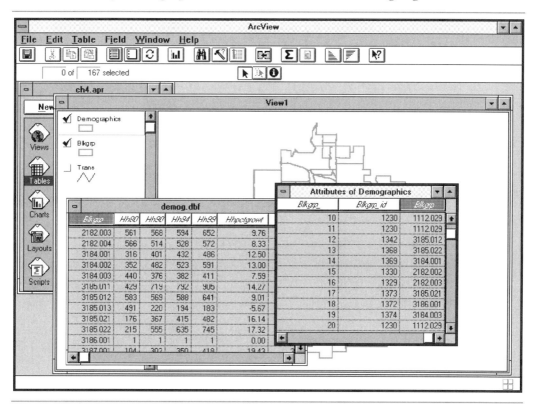

The Attributes of Demographics and demog.dbf tables with the highlighted join field, Blkgrp.

6. With both fields highlighted, select Join from the Table pull-down menu on the menu bar. The two tables are now joined on the common field. A status bar at the bottom of the application window displays progress.

7. When the join is complete, the *demog.dbf* table is closed and the *Attributes of Demographics* table contains the attributes from the *demog.dbf* table. Scroll through the Attributes table to view the results of the join. When you have finished examining the joined table, close the window displaying the table.

➣ *NOTE: Order is important when joining tables. As mentioned earlier in this chapter, join requires a one-to-one or a many-to-one relationship. When working with theme attribute tables, the theme attribute is the primary feature of interest. Accordingly, the theme attribute table should always be the destination table. In selecting the fields on which to join, we selected the field from the text file first and the attribute table for the theme second to ensure that the attribute table window was the active window. When active, the window for the active table is highlighted. Making this window active ensures that the Attributes of Demographics table will be the primary table during the join. After the join, the fields from the secondary table are appended to the primary table. Joining these fields to the attribute table for the theme makes the fields available for subsequent analysis and mapping.*

At this point, we want to perform the same operation for a second table titled *cable.dbf*. The steps for the second join follow.

1. Copy the *Blkgrp* theme using the Copy/Paste procedure described earlier. Rename the theme *Cable*.
2. Open the attribute table for the *Cable* theme, and open the *cable.dbf* table.
3. Join the attribute and *cable.dbf* tables on the *Blkgrp* item. (The incremental project has been saved as *ch4a.apr*.)

We have now linked two of our three attribute tables (*demographic* and *cable* tables) to a spatial theme (*Blkgrp*) so that the data can be mapped. Our third file, *restrnt.dbf*, is a little different. In addition to fields containing the business name and SIC Code for each restaurant, *restrnt.dbf* contains two fields, the x and y coordinates, which are latitude/longitude points identifying restaurant locations. The x and y coordinates allow for rapid geocoding.

Exercise 3: Extending Project Data 85

In order to map the restaurant locations in *restrnt.dbf* using the coordinate values, we need to transform the table into an event theme. The following steps outline this task.

1. Select Add Event Theme from the View pull-down on the menu bar.
2. In the Add Event Theme dialog window, click on the XY icon to specify adding an x,y event theme.
3. Select the *restrnt.dbf* table from the table scroll list.
4. Specify the fields in the table to use for x and y coordinates in the X Field and Y Field scroll lists. Select Long and Lat. Click OK.

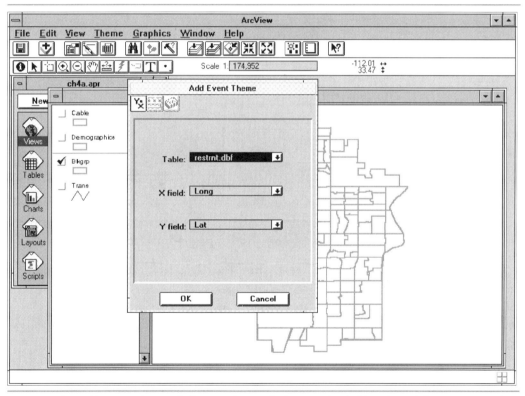

The Add Event Theme dialog window with the X and Y fields selected from restrnt.dbf.

5. The completed event theme is added to the legend. A default point symbol and color is assigned, which you can subsequently change using the Legend Editor. Click on the box associated with this theme to display it in the view. You will see a number of points,

Chapter 4: Extending Data

most located along a grid corresponding to the arterials in our study area. (The incremental project has been saved as *ch4b.apr*.)

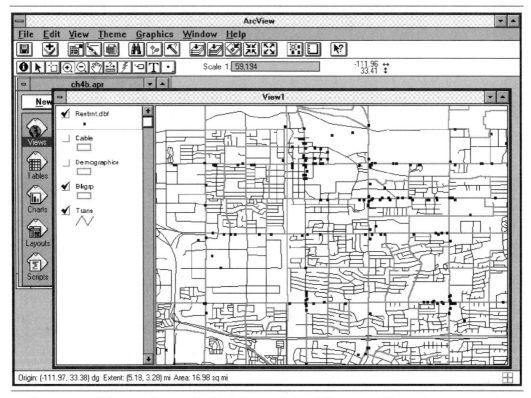

The restrnt.dbf event theme displayed against the Blkgrp and Trans themes.

With the demographic data and restaurant locations as background, we now want to direct attention to our client base. As you may recall, the purpose of the exercise is to demonstrate how ArcView can support market research. (See insert titled "Primary Research.") Our overall program is described below.

We worked with a local restaurateur to design a snapshot survey that would capture information about restaurant clientele. In addition to customer home addresses, we asked for additional information on preferences. Two short questionnaires were developed: one focused on potential advertising strategies to reach the clientele, and the other on assessing the client versus the competition.

Exercise 3: Extending Project Data   87

We received 67 responses of each form for a total of 134. It should be noted that the number of usable responses (111) was less than the sample size originally targeted when the survey was designed. Statistical formulas indicated that a sample size of at least 300 was required for the level of accuracy desired in evaluating our customer base. While the conclusions we derive from this data will not be statistically significant, the data serve to illustrate the overall process.

To add survey information to the project, take the following steps:

1. From the Project window, select Add Table from the Project pull-down menu.

2. Switch to the *$IAPATH\data* directory, and add the *ihopad.dbf* and *ihopcomp.dbf* tables. These tables contain the responses from the two customer surveys.

    A quick browse of the records in these tables will reveal that while the restaurant is located in Tempe, customers come from all over the valley. For the purposes of this exercise, we will confine our analysis to respondents located in Tempe.

3. Make each table active by clicking on the title bar.

4. Select Query from the Table menu (or click on Query Builder tool). Select records for which *City = Tempe*. (*City = Tem* for *ihopcomp.dbf*.) The resulting selected set will be 30 of 67 for *ihopad.dbf*, and 19 of 67 for *ihopcomp.dbf*.

[handwritten note: DOUBLE CLICK ON CITY / CLICK ON = / DOUBLE CLICK ON TEMPE]

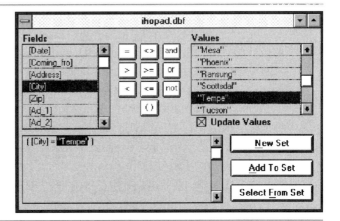

Selecting City = Tempe using the Query Builder.

The next task is to create a theme from each *dbf* table with a point located at each record's street address. In ArcView terms, we will create a

88   Chapter 4: Extending Data

geocoded event theme. The following steps are the same ones used to locate the day care centers in Exercise 1 (Chapter 2).

1. Click on the *Trans* theme in the Table of Contents to make the theme active.
2. From the Theme Properties dialog window, select the Geocoding icon from the scroll list on the left. Accept all default address field choices. Click on OK to build the geocoding street index on the *Trans* theme.

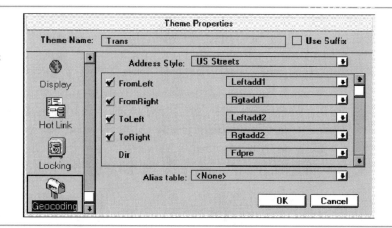

The Geocoding Theme Properties window for the Trans theme.

3. When the geocoding street index is complete, you are ready to add the event themes. From the View pull-down menu, select Add Event Theme.
4. Select the right icon (the letter) to add a geocoded theme; select *ihop.dbf* from the list of tables, and accept Address as the field to use for geocoding. Accept <none> for the Join Table and Alias Table because we will not be physically joining the two tables, nor will we use an alias table for address matching.
5. At this juncture, you will be presented with the Geocoded Theme Name dialog window. Switch to your working directory and enter *ihop1a.shp* for the theme name. Click on OK.
6. Now you see the Geocoding Editor dialog window. The first match is presented, along with the option to step through the records manually by selecting the Match button, or to automatically match the entire file by using the Start button. Select the Start button.

Exercise 3: Extending Project Data   89

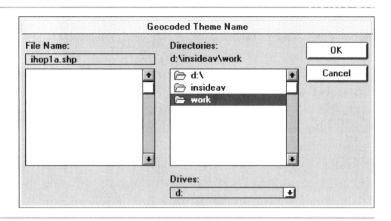

Naming the ihop1a.shp theme in the Geocoded Theme Name dialog window.

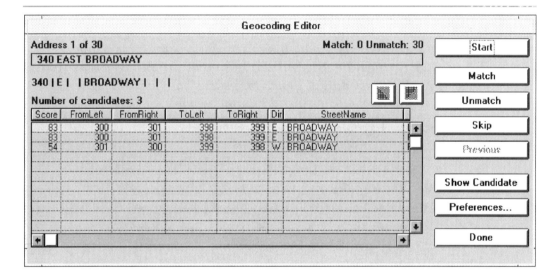

The first record in the Geocoding Editor for ihop1a.dbf.

7. At the completion of address matching, the status line will show 21 records matched, and 9 unmatched. Click on Done to accept the results; the *ihop1a.shp* theme is added to the Table of Contents.

8. Follow the same procedure to create a geocoded event theme from *ihopcomp.dbf*. For the output file name, use *ihop1c.shp*. Upon completion of address matching, 11 records will be matched and 8 unmatched.

As you may have noticed, the match rates for these tables were not particularly high. Because we are working with a small sample, it is important that we match as many records as possible. A quick check of the *ihopad.dbf* and *ihopcomp.dbf* tables reveals several records with incorrect formatting, such as *ASU Registrar's Office* and *Priest // University*. With the aid of a street atlas, we are confident we can obtain a higher match rate. We could match the table again by clicking on the *ihop1a.shp* theme in the Table of Contents to make it active, and then select Rematch from the Theme pull-down menu. This time we can elect to match the records one by one by stepping through them using the Match button. In this manner we can interactively edit any record which does not match to correct format, spelling or address errors, and resubmit the address match process following the edit.

> **NOTE:** *Interactive address editing using the Geocoding Editor only affects the address used for matching in creation of the event theme. It does NOT change the original address value in the associated table. The address used by ArcView for geocoding the event theme is stored in a separate field, Av_add, in the theme's attribute table.*

For this exercise, we have carried out the editing for you. To geocode the edited tables, take the following steps:

1. From the Project window, switch to the *$IAPATH\data* directory and add the *ihopad2.dbf* and *ihopcmp2.dbf* tables.
2. Geocode the tables as demonstrated above. Remember to select out the Tempe records. For output names, use *ihop2a.shp* and *ihop2c.shp*. This time, when the geocoding is complete, you will obtain a 100 percent match.
3. Display the new geocoded event themes to view the newly added points. Select and delete the old event themes because we will no longer need them. Rename the new event themes by accessing the Theme Properties dialog window for the active theme. Rename the *ihop2a.shp* theme to *IHOP - AD*, and the *ihop2c.shp* theme to *IHOP - COMP*.

We are now at the point where analysis begins. In Chapter 5, we discuss querying and classifying data, the first phase toward obtaining an initial overview of what the data represents.

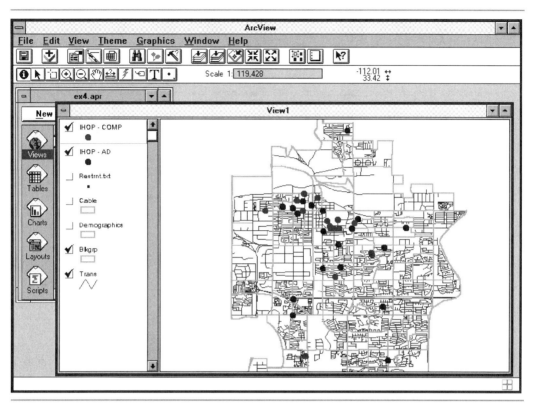

The completed IHOP - AD and IHOP - COMP event themes.

Primary Research

To solidify your thinking on how we might use ArcView in a real world setting, in the chapters ahead we will conduct market research for a restaurateur from Tempe, Arizona. In this ongoing example, we will demonstrate how ArcView can be used as part of a research project.

As regards background, for many years retailers and others dealing in products serving the general public have conducted *pin* studies (in reference to the map pins originally used to locate responses) to identify where their customers reside or work. By knowing where their customers come from, "trade areas" can be established and subsequently linked to demographic data to further define customers. This process can aid in designing future marketing strategy, such as direct mail or outdoor advertising. Where franchising is involved, pin studies are also used to define "exclusive rights" territories.

The two major steps involved in conducting a pin study are capturing customer addresses and transferring the customer location data to maps.

Data is typically collected via a customer intercept or "fishbowl" survey. Patrons are asked to fill out a form describing themselves and drop the form into a jar as they leave the business premises.

The data set provided in our exercise, *cust.dat*, was collected through an actual fishbowl survey at a family restaurant in Tempe. The address matching capability of ArcView will be used to expedite the traditionally tedious task of locating and analyzing customer records.

Once customers are located, their distribution in space can be mapped. By linking this data to commercial "profiling" data, we can also study what type of customers the restaurant or other business is reaching. Profiling data consists of a wide range of parameters that identify who the customers are and how best to reach them. Examples of these parameters are the TV or cable channels they watch regularly, and magazines they subscribe to.

The GIS approach through ArcView dramatically improves the analysis of pin studies. Data is processed much more efficiently, and the modeling and analytical tools allow the user to reach much more powerful conclusions.

Chapter 5

Displaying Data

In Chapter 3 we imported tables and spatial data, and in Chapter 4 the data was extended by joining tables and creating event themes. At this juncture, the data is ready for display. Although we have touched upon some display issues in previous exercises, in this chapter concepts involved in data display will be explored in greater depth.

Defining Symbology

When a theme is added to a view, ArcView assigns it a default symbol, and the symbol color is randomly selected. Because you will often wish to override the system-assigned choice, ArcView makes it easy to change symbols.

The first step in changing symbols is to access the Legend Editor window by double-clicking on the theme entry in the Table of Contents. If the theme is already active, you can call up the Legend Editor by selecting Edit Legend from the Theme menu or by clicking on the Legend Editor icon from the button bar.

✔ *TIP: If the Legend Editor window is already open, double-clicking on the theme entry in the Table of Contents will reinitialize the Legend Editor for the current theme.*

There are several display options in the Legend Editor. With the editor you can change a theme's symbology or classify a theme. Classification will be discussed later.

When a theme is initially added to a view, one symbol is assigned to draw the entire theme. When brought into the Legend Editor, the symbol

94 Chapter 5: Displaying Data

for the theme is displayed, and the area for display of legend text is blank. To change the symbol for theme display, double-click on the symbol; the symbol palette window will be displayed.

The functional areas accessed through the symbol palette window are indicated by six icons displayed across the top of the window. From left to right these icons access the Fill Palette, Pen Palette, Marker Palette, Font Palette, Color Palette, and Palette Manager windows.

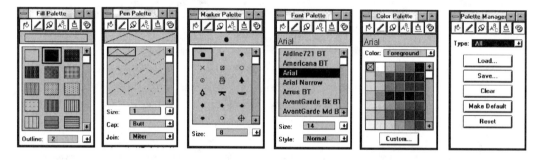

The Fill Palette, Pen Palette, Marker Palette, Font Palette, Color Palette, and Palette Manager windows.

The Fill Palette controls how areas are shaded. Through this window you can indicate whether the corresponding polygons should be shaded in full, with hatchings or not at all. In addition you can control polygon outlines, or border width and appearance.

> **NOTE:** *Polygon themes consisting of relatively small polygons covering a large area, such as land use or major land ownership, are better shaded without polygon outlines, particularly when displayed at large scales.*

The Pen Palette controls the appearance of line features and line themes. You can adjust the pattern and width of lines, as well as settings for how ArcView draws the ends of lines and line vertices through the Cap and Join options.

The Marker Palette controls the appearance of point features and point themes through the ability to change a point's symbology and size.

The Font Palette controls how text and labeling will appear. Available options include font type, size, and style (normal, bold, italic, and bold italic). On the Windows platform, all TrueType fonts installed on your system are available.

The Color Palette controls color. Whether you are working with areas, pens, markers or fonts, this palette allows you to control a feature's foreground color, background color, outline and text. In addition to the standard color palette, a custom option is available to allow for color mixing by specifying Hue, Saturation and Value.

Finally, the Palette Manager allows you to import new symbol palettes or revert to the default palette.

ArcView palettes enable you to control every aspect of how your features appear. As you gain skill in customizing maps, their use will become second nature.

Classification

We have thus far addressed changes to theme symbology where one symbol is used to represent the entire theme. The next step is to *classify* the theme based on the values of an attribute field associated with the theme. Classification by attribute value is the cornerstone of thematic mapping, and GIS as well. Through this single function you will open up a wealth of information pertaining to *thematic data.* By providing fast and easy access to the tools for thematic classification, ArcView allows you to quickly analyze the patterns underlying thematic data.

The first step in thematic classification is to access the Legend Editor by double-clicking on the theme entry in the Table of Contents. The user must then choose the field upon which to classify. Access the Field pull-down menu and click on the name of the field you wish to classify. Classification will now begin.

> *NOTE: Fields on the theme's primary table, and fields from joined tables are all classifiable. Any field which has been joined to the theme's attribute table can be classified and mapped.*

When classification is complete, a default five-class grouping is displayed; each class contains an equal number of records. You have the option of further adjusting the classification type, the number of classes or value of specific classes. These options are described below.

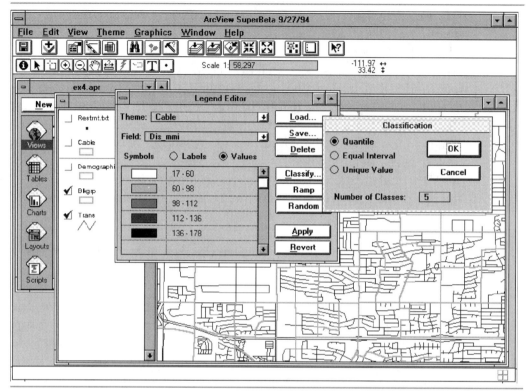

The Legend Editor and Classification windows.

Classification Type

The following classification methods are available from the Classification dialog box:

❒ Quantile (default). Classes contain an equal number of records.

❒ Equal Interval. The resulting classification contains an equal range of values in each class.

❒ Unique Value. Each unique value is placed in a separate class. The maximum number of unique values is 64.

In addition to the choices seen in the Classification dialog box, a previously saved *user-defined* classification may be loaded and applied to a new theme, providing that the class breaks are applicable to the new theme. The user-defined scheme in such cases is added after the initial classification (one of the three listed above) is performed. To apply a previously saved classification, you would use the Load button in the Legend Editor dialog window.

Number of Classes

For quantile and equal interval classifications, the user can specify the number of classes ranging from two to 64.

Class Values

A class range can be modified by clicking on the Values button and keying in new values for class ranges. The revised classes can then be applied to the theme. Modifying class ranges is a method by which users define the classification scheme.

> **NOTE:** *ArcView does NOT check to verify whether you have entered valid class ranges. If class ranges overlap, ArcView will assign the feature to the first class into which the feature's attribute falls. In the event of gaps between classes, features falling in this range will not be drawn.*

When you are satisfied with the classification scheme, the next step is to arrange how the new classes will be displayed. Double-clicking on individual ranges in the legend will call up the symbol palettes as described above. The symbol and color assigned to each class can be customized using this method.

Random and Ramp Tools

The Random and Ramp tools are used for assigning color to classes. The Random button assigns random colors to the classes. This feature can serve as a quick starting point for evaluating a tentative classification.

The Ramp button is used to create a color ramp, or color gradations, between the lowest and highest classes. In this method, the user assigns a color value to the lowest and highest classes. Clicking on the Ramp button will generate the color ramp for the intermediate classes. This option is particularly valuable when classifying continuous data, such as population or age.

Classification Revisited

Classification methods can have a profound effect on how data will be displayed on a map. The selection of *quantile*, *equal interval*, or *unique values*, combined with different total class numbers, may produce strikingly different results.

Let's consider a small data set of 12 values: 24, 25, 26, 29, 32, 43, 44, 51, 69, 78, and 113.

Nearly any method of automatic ranging (quantile or equal interval) will skew the results toward some type of bias. Using four classes, the quantile approach will set breakpoints at 26, 43, and 51 in a manner that arbitrarily splits and artificially communicates breaks in the natural progression between 26 and 32 and 43 and 44. Using the equal interval approach, the breakpoints will be set at 46, 69, and 92. Again, the breaks seem less than optimal: this time, 49 and 51 are unnecessarily separated from the lower group, and 69 and 78 are split apart.

Could this situation be resolved with fewer classes? Experimenting in this manner is healthy, but not guaranteed to be productive. Using the same numbers, three quantiles produce breakpoints at 29 and 49, again resulting in seemingly artificial splits between values in very close proximity. However, three classes, with breaks at 54 and 84, might work. This group, more than any of the others above, represents the actual distribution of the data.

Is such careful work with ranges typical? Actually, it is because data rarely fall into a distribution where a quantile or equal interval classification is optimal. For many data sets, automatic classifications might not be the answer. In these situations, you should feel free to explore custom breakpoints. One way to proceed is to sort the table and visually determine the natural breaks. Another is to utilize ArcView's statistical capabilities.

ArcView can provide summary statistics for a table based on the active field. Available statistics include average, sum, minimum, maximum, standard deviation, first, last, and count. These statistics can be generated for any numeric field, and are written to a new table. The table can subsequently be joined to the Theme Attribute table and used for additional analysis.

Why is the careful choice of ranges so important? Because ranges will be used to determine the color breaks in your map, users will ultimately remember the colors better than the specific ranges that the colors represent.

In *How to Lie with Maps* (The University of Chicago Press, 1991), Mark Monmonier examines how reality can be distorted by the way maps are designed. We recommend that you seriously consider the issues mentioned above; otherwise, your maps could end up being more misleading than helpful.

Cleaning Up the Legend

Once the symbology has been set, usually the only task remaining is to edit the legend text. This, too, is accomplished from the Legend Editor window.

To edit the legend text for the classification, click on the Labels button in the Legend Editor window. Highlight the text string for the first class by clicking on it. Key in the new legend text, and <Enter>. The next text string will be highlighted. Text for the entire legend can be edited in this manner. If you wish to leave a legend text string unchanged, press <Enter>. If you wish to return to the original legend labels, close the window or select Revert.

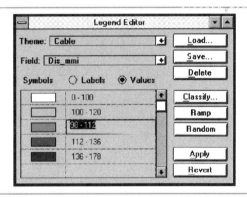

Entering new class ranges into the Legend Editor.

When all desired changes have been made, clicking on the Apply button will cause them to be applied to the theme. At any point before clicking Apply, clicking on Revert will reset the theme to its initial state.

Hiding the Legend

By default, when a classification is applied to a theme, the theme legend is displayed in the Table of Contents. However, legends can be quite extensive, particularly those resulting from a classification using unique values. As themes are added to a view, it does not take long before the Table of Contents is too long for all entries to be displayed on the screen. Thus, you need to scroll through the Table of Contents in order to view all theme entries. To reclaim space in the Table of Contents, you have the option of hiding or showing the classified legend for each theme. When the legend is hidden, the theme will be represented with just the theme title and check box in the Table of Contents. This feature is accessed by the Hide/Show Legend choice from the Theme menu of the View menu bar.

Identifying Features

At this point, and continuing through Chapter 6, we will look more closely at examining the data associated with spatial themes. One straightforward way to examine the attributes of a theme is with the Identify tool, located at the far left of the tool bar. The Identify tool allows you to display the attributes of a feature in an active theme. To make a theme active, click on its entry in the Table of Contents.

> ✔ **TIP:** *When you click on a theme name while simultaneously pressing <Shift>, the theme becomes active along with all previously active themes.*

With the Identify tool active, clicking on a feature in the view brings forth a window displaying feature attributes. Identify displays each column and the corresponding values for the identified feature.

The Identify tool icon.

A list is maintained on the left side of the window of all identified features. If more than one feature was found at that location, multiple records will be displayed in the list. Clicking on a record from the list will refresh the record's fields display. As additional features are identified, they are added to the list. Consequently, current and previous Identify results can be compared.

If two or more themes are active, a record will be added to the list for each feature found from each active theme. This is a simple way to compare values for overlapping themes at a specific location.

The Clear button clears the record for the current Identify, while Clear All removes all records from the Identify report.

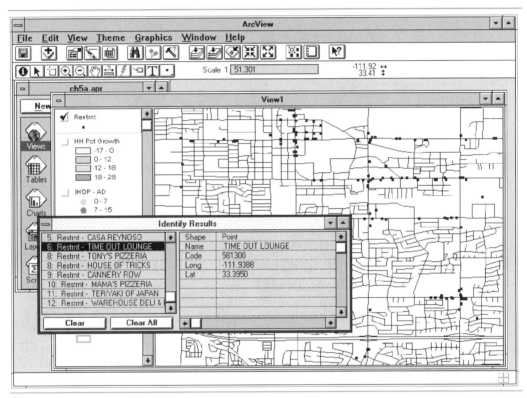

Results of Identify on the restaurant theme.

Labeling Features

Another way to identify features in a theme is with the Label tool. The Label tool, located on the tool bar, allows you to select a feature from the active theme which is subsequently labeled with the value from a designated field.

> **NOTE:** *Unlike Identify, Label displays the value for only one field from a single active theme.*

To specify the field for feature labeling, access the Theme Properties dialog window from the Theme pull-down menu. Click on the Text Labels icon to access the text labeling options. A pull-down menu displays the available choices for Label Field, as well as options for how to orient the text.

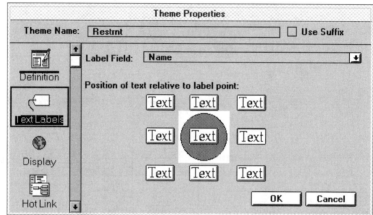

The Text Labels Theme Properties window.

In order to label features, the theme must be active, and the theme must currently be displayed. With the Label tool active, click on a feature in the active theme. As you select features, a text label is placed adjacent to the feature containing the selected attribute from the Label Field.

The resulting text labels are ArcView graphics elements, and can be moved, resized, and edited in the same manner as any other graphic element. In addition, the text labels are attached to the theme by default. The elements are turned off when the theme is off, and if the theme is subsequently deleted, the attached text labels will be deleted as well.

It is also possible to detach text labels from a theme. To do so, make the theme active and select Detach Graphics from the Theme pull-down menu.

Adding Graphics to a View

Text labels are only one type of graphic element which can be created in ArcView. Drawing tools also enable you to create points, lines, polylines, rectangles, circles, and polygons. These graphic elements can be added to a view or a map layout. When added to a view they can be linked to a theme, or used to spatially select features from the theme.

Adding Graphics

The Draw tool is used to add all graphic elements, with the exception of text. When you press the left mouse button on the Draw tool icon, several icons appear which correspond to the graphic types listed above. Graphic elements are added by clicking with the mouse as described below.

- ***Points*** are located with a single mouse click.

- ***Lines*** are graphic elements which contain a start and an end point. The start point is identified by clicking; the mouse button is held while the mouse is dragged to the end point, where it is released. The length of the line is displayed as the line is dragged.

- ***Polylines*** are lines which contain more than two points. To add a polyline, click at the start point and at the location of all intermediate points. Double-click at the end point to end the polyline. The length of the last line segment and cumulative line length are displayed as the polyline is drawn.

- ***Rectangles*** are added by clicking at the start point, pressing the mouse button, and dragging to define the box. The display shows the current area of the rectangle as the box is dragged.

- ***Circles*** are added by clicking at the center, and then pressing the mouse button and dragging to define the radius. The display shows the current radius as the circle is dragged.

- ***Polygons*** are added by clicking at the start point and subsequent vertices. Double-click at the last vertex to close the polygon. The display shows the last segment length, perimeter, and area as the polygon is defined.

- ***NOTE:*** *The units of length or area displayed as graphic features are set via the Distance Units selector from the View Properties dialog box. In addition, the Measure tool from the View tool bar can be used to measure distance without adding a polyline to the view.*

Text is added to a view by using the Text tool from the tool bar. With the Text tool active, click at a point in the view to anchor the text. The Text Properties dialog window is then opened. The window includes an area for text input, as well as options for line justification, vertical spacing between text lines, and rotation angle for the text. After the text is entered and properties set, click on OK to add the text to the view.

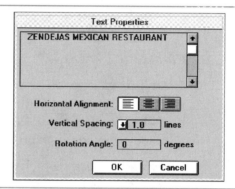

The Text Properties dialog window.

Graphics elements added to the view use the current settings from the Symbol Palette. To access the Symbol Palette, choose Show Symbol Palette from the Window pull-down menu, or key in <Ctrl>+j.

Editing Graphics

Graphics elements can also be moved or edited after they have been added to a view. To select a graphic element, use the Pointer tool on the tool bar.

The Pointer tool icon.

Individual graphic elements are selected by clicking on them with the mouse. Several elements may be simultaneously selected by dragging a selection box with the mouse. To add additional elements to the selected set, hold down the <Shift> key while clicking on additional elements.

To delete selected elements, choose Delete Graphics from the Edit menu, or use the <Delete> key after the elements are selected. In the same manner, elements can be copied, grouped, or moved between views.

When the graphic element is selected, the selection handles for the element will be displayed. To move an element, place the cursor inside the element's

Adding Graphics to a View 105

boundary box. The appearance of the cursor will change to crossed arrows. Press on the left mouse button and drag the element to a new location. If more than one element is selected, all elements will be moved as a group.

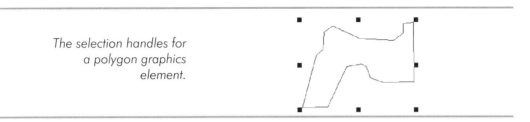

The selection handles for a polygon graphics element.

To resize elements, drag on one of the selection handles. Dragging on a corner handle will resize the element proportionally. Dragging on a side element will stretch the element in that direction.

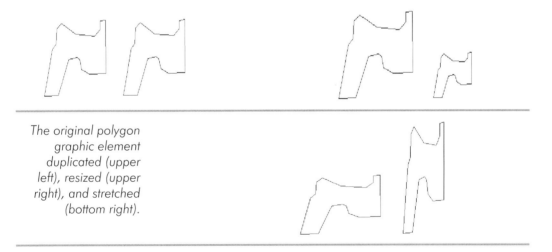

The original polygon graphic element duplicated (upper left), resized (upper right), and stretched (bottom right).

To change the display properties of an element after it has been added, select the element or elements and access the Symbol Palette. Symbol and color choices will be applied as they are selected.

Reshaping

In addition to moving graphics, you may wish to occasionally edit their shape. To reshape polylines and polygons, click on the graphic element to reveal its selection handles. Clicking on the element again will cause the selection handles to be replaced by the vertex handles. Click and drag on the vertex handle to reshape the graphic by moving the vertex to a new location.

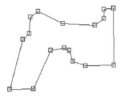

The vertex handles of a polygon graphic element.

Text Editing

Text labels may be resized and repositioned using the Pointer tool. In addition, text labels are edited by clicking on the text label with the Text Tool active. The text element will be displayed in the Text Properties window; the text can then be edited and additional properties changed as needed.

Exercise 4: Graphics, Symbols and Classification

Open the project you saved at the end of Exercise 3. If you did not save the project, open the *ch5.apr* project file in your working directory.

Turn off the *IHOP* point themes, and the *Blkgrp* and *Trans* themes. Turn on the *Demographics* theme, and click on the theme entry in the Table of Contents to make it active.

From the Edit pull-down menu, select Copy Theme, and then select Paste Theme from the same menu. A copy of the *Demographics* theme will appear at the head of the Table of Contents. Click on this new theme to make it active.

The *Demographics* theme contains a variety of demographic data, including information on population and income. We will examine some of these attributes in greater detail, starting with the field containing the projected percent growth in households from 1994 to 1999.

Exercise 4: Graphics, Symbols and Classification 107

Access the Theme Properties dialog window and change the name for this new theme from *Demographics* to *HH Pct Growth*. Next, double-click on the theme in the Table of Contents to bring up the Legend Editor. From the Fields list, select *Hhpctgrowth*. A five-class quantile classification is immediately generated. Apply this classification, and examine the results.

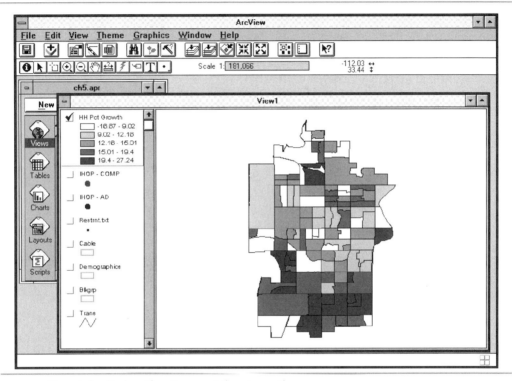

The default classification on Hhpctgrowth.

The default symbolization is a solid-fill gray scale from white to black. Changing the symbol colors may aid in determining if this is a useful classification. Because we are working with continuous data, the color ramp is appropriate.

Double-click on the symbol for the first class to bring up the Symbol Palette window. Switch to the Color Palette, and select a pale color. Now, click on the highest class, and select a color with strong saturation. Click on Ramp.

ArcView constructs a color ramp grading between the two colors. In some cases, all the intermediate colors are readily distinguishable and

Chapter 5: Displaying Data

easily interpreted as representing classes of continuous data. In other cases, one or more colors may be difficult to distinguish. Adjust your selections until you get a ramp you like. Next, click on Apply to apply these symbols to the classified theme. If you cannot find a color selection that *ramps* well, you might try a ramp between pale blue and a strong orange.

With the new colors applied, the results of the classification should be more evident. A pattern is discernible, but assume we wish to modify the class breaks to further explore the data. To change class breaks, click on the text for the legend entry of the first class. The text will be highlighted. For the first class, key in the range -17–0 (negative 17 to zero). When you press <Enter>, the cursor will be automatically positioned at the next field. Continue down the list, and key in the following class ranges: 0–10, 10–15, 15–20, and 20–28. When the class range changes are complete, click on Apply. The new classes are immediately applied, and the view display is updated.

➜ ***NOTE:*** *It is customary to repeat the value for the upper range of a class as the lower range of the next class. If a field value falls on the break, it is assigned to the first class containing that value. Thus, the value 10 would be assigned to the 0 - 10 class.*

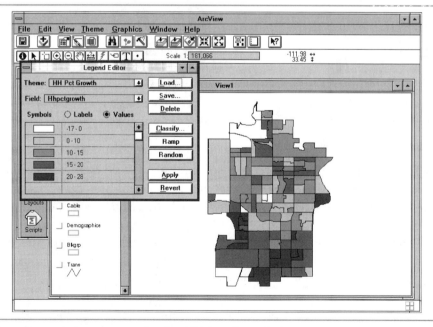

The adjusted class ranges for the HH Pct Growth theme.

Exercise 4: Graphics, Symbols and Classification   109

We can likely obtain an even better pattern if we reduce the number of classes. In the Legend Editor dialog window, click on the Classify button, and change the number of classes from 5 to 4. A new four-part classification is generated. Note that our previous color choices are preserved. Edit the class ranges again, this time using the following ranges: -17 - 0, 0 - 12, 12 - 18, and 18 - 28. Apply the classification, and examine the results.

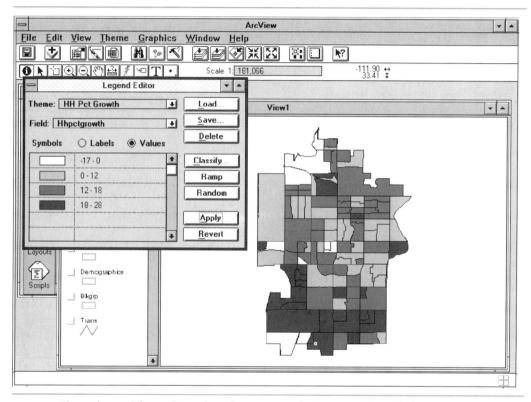

The adjusted four-class classification on the HH Pct Growth theme.

The four-part breakdown appears to adequately portray the data breaks. Finally, click on the Labels button in the Legend Editor, and key in the same values for the new class ranges. Applying this change will cause the legend for this theme in the Table of Contents to be updated with the new values. You can also substitute "Less than 0" and "Greater than 18" for greater clarity. (The incremental project has been saved as *ch5a.apr*.)

Next, let's examine the amount spent item in the restaurant customer survey. We are going to map this item in an attempt to identify patterns.

110   *Chapter 5: Displaying Data*

1. Double-click on the *IHOP - AD* theme to access the Legend Editor. From the field list, select Spent.
2. The default quantile classification is applied. A cursory look at the class breaks suggests that five classes may be excessive. Click on the Classify button, and change the number of classes from 5 to 3. The class breaks are now 0–7, 7–15, and 15–30. Change the marker colors as desired to better distinguish the three classes. Click on Apply.

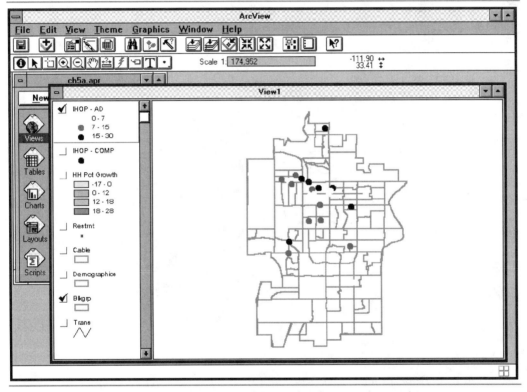

The three-class classification on amount spent for the IHOP - AD theme.

3. Perform a similar type of adjustment to the *IHOP - COMP* theme. Adjust the class breaks so that they correspond to those in the *IHOP - AD* theme. Assign the same symbol colors, and click on Apply.

Exercise 4: Graphics, Symbols and Classification 111

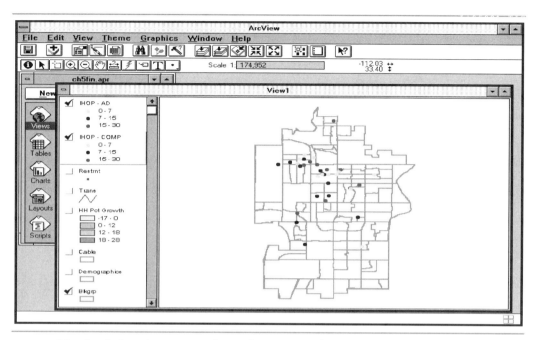

The final classification on the IHOP - AD and IHOP - COMP themes.

In the display, dot size was made smaller by changing the size from 8 to 6 via the Marker Palette. To change dot size after classifying the theme, double-click on the symbol for the first class to re-initialize the Palette Window, and select the Marker icon (third from the left) to access the Marker Palette. With the <Shift> key depressed, click on the remaining class symbols until all are highlighted. Click on the arrow to the right of the size input box to access the scrolling list of available marker sizes. While the scrolling list jumps from 4 to 8, it is possible to backspace over and enter a custom value for size. In this instance, enter 6. Apply the new marker size to the classified theme.

For the final display, turn on the *HH Pct Growth* theme, and display the IHOP survey points against this theme. If you are starting to see patterns emerging, and are considering the application of additional classifications, you are beginning to appreciate the power of classification in GIS analysis. (The incremental project has been saved as *ch5fin.apr.*)

112   Chapter 5: Displaying Data

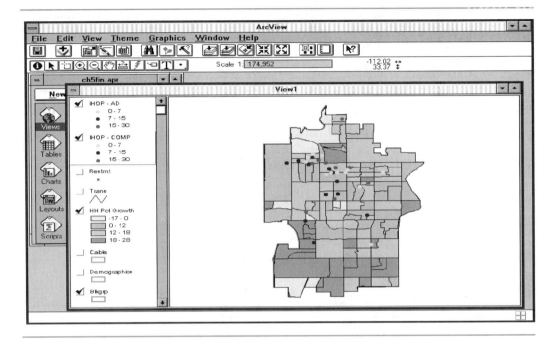

The IHOP - AD and IHOP - COMP themes displayed against the HH Pct Growth theme.

The Identify and Labels tools are fairly straightforward. We suggest that you turn on an active theme and identify and label features until you feel comfortable. The same goes for the graphics tools. Turn off enough themes so that your display is relatively uncluttered, and start experimenting with adding different types of graphics. Switch to the Pointer tool to select graphics, and move, resize, stretch and reshape them until you feel comfortable with these manipulations. Select and manipulate a few text labels as well. Mastering ArcView graphics is particularly useful for performing spatial manipulations on themes.

Chapter 6

Data Queries

This chapter is focused on manipulating tables as an analytical tool. In preceding chapters, attribute tables associated with x,y and address geocoded event themes were discussed. Tables were imported from delimited text files and joined to theme attribute tables. Through all of the above operations, table appearance did not change.

From the ArcView user's perspective, all tables *are the same*, regardless of table source (INFO, dBase or delimited text file). ArcView defines a standard template to reference the tables you access. The tabular data itself is *not* imported, but rather continues stored in the source file in native format. The ArcView link to the data is *dynamic*: changes in your data outside ArcView will be reflected in ArcView projects that reference the data. Consequently, ArcView sees the same snapshot of your data as other application packages. There is no redundancy because you do not manage multiple databases. Your database software handles your tabular data, freeing ArcView to focus on managing and organizing spatial data.

Basic Table Operations

Upon opening a table, all fields are displayed along with the field names from the initial table definition. This may or may not be satisfactory depending on the size of your table and how the fields are named. Tables containing many fields require scrolling of the display to examine widely separated fields. Field names may be incomprehensible, making interpretation difficult. Fortunately, these properties can be changed.

The Table Properties window controls which fields are displayed as well as an *alias* for each field. An alias allows the user to substitute a more

Chapter 6: Data Queries

easily recognizable name for a cryptic field name. For example, a field defined as *VLPCT* can be aliased as "% Vacant Land." The new name will then be used in future queries and output. To access the Table Properties window, click on the title bar of the table to make it active, and then select Table Properties from the Table pull-down menu.

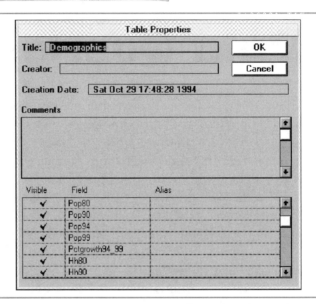

The Table Properties window.

A list of all fields in the table is displayed, along with a field to specify the alias name for the field, and a check field indicating whether the field will be visible in the table. To change the alias name for the table, click on the Alias cell for the field, and type in the new name. This name will now be associated with this field in all subsequent operations.

The Visible field controls which fields are displayed in the table. Initially, all fields are set to be visible. To make a field invisible, click on the Visible cell for that field. This will toggle off the check mark rendering the field invisible.

> **NOTE:** *Hidden fields are not available for query or classification. They will not be shown in the results of Identify nor in feature labeling, and will not be included when you print or export the table.*

In addition to making a field visible or invisible, fields can also be resized and reordered on the display. To resize a field, place the cursor on the border of the field name at the head of the table. Click and drag the border to increase or decrease the display width for the field. If desired, the display width can be collapsed so that the field does not display; the field will still be available for all other operations.

To change the order of the fields in a table, click and drag the name of the field at the head of the table. As you do so, the outline of the cell will be displayed. When you have positioned the field where you want it, release the mouse button. The table will be redisplayed with the field in the new location.

Sorting a table may also make it easier to locate specific records in that table. To sort a table, click on the field name on which to sort, and then select the Sort Ascending or Sort Descending button from the button bar. The table will be redisplayed in sorted order.

The Sort Ascending and Sort Descending icons.

Using Tables with Views

Once you have added themes to a view, opened the attribute tables associated with these themes, and joined tables as needed, the next step is to associate records from these tables with features on the view. These operations can range from simple selection and identification of features to robust logical queries.

Selecting Features

One of ArcView's basic functions is the ability to select features. As features in the active theme or records in the active table are selected, they are highlighted in a different color. This color, which defaults to yellow when ArcView is initialized, can be set from the Properties dialog box of the Project window. When the attribute table is open for the active theme, the selected features are highlighted both on the view and in the table.

The simplest way to select features is with the mouse. Clicking on a feature from the active theme causes it to be selected, and drawn in the current highlight color. The corresponding record in the attribute table for the active theme will be highlighted as well. As an alternative, you can keep the mouse button depressed to allow multiple features to be selected by dragging a box over the desired area.

Clicking on a new feature causes it to replace the prior feature as the selected feature. To add to the selected set, hold down the <Shift> key while selecting additional features. Clicking on a selected feature with the <Shift> key depressed will cause the feature to be removed from the selected set.

In the same manner, features can be selected by clicking on a record in the attribute table for the active theme. As records are selected in the table, the corresponding features are highlighted in the view.

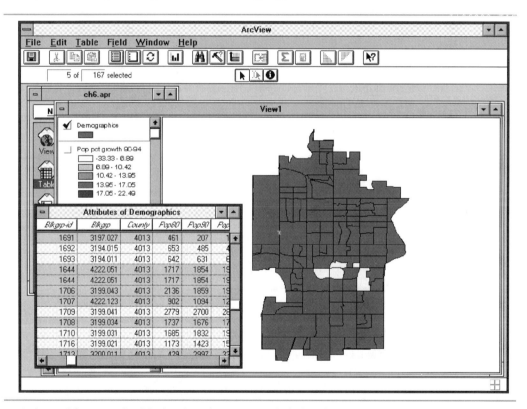

Selected features highlighted in the view and in the theme attribute table.

Selecting Features with Shapes

To more easily view the records for the selected set, the Promote tool from the Table button bar reorders the table and displays the selected records at the top of the table.

The Promote icon.

Selecting Features with Shapes

The ability to select features by dragging a box with the mouse as described above is but a sample of the ability to select features by shape.

Selecting features by shape is a two-step process. First, create the graphics shape using the Draw tool on the View tool bar. Second, use the resultant shape or shapes to select features from the active themes. Existing graphics may be used as well as graphics created specifically for selecting features.

The Select Features Using Shape tool on the View button bar selects all features in the active themes falling partially or totally under the selected graphics. These graphics can be a single shape, such as a line or polygon, or a mix of shapes, such as circles and polylines.

The Select Features Using Shape tool.

The ability to select features using shapes lends itself to basic analytical techniques. A circle can be created with a specific radius (e.g., one mile) to delineate areas of influence for point locations. Polygons can also be drawn to represent service areas, and used to select businesses or homes lying within this area.

 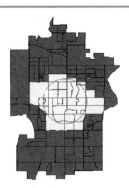

A circle drawn on a view (left); the features selected using the circle (right).

Selecting Features by Query

Occasionally, you will want to identify a feature by its value. In these cases, you do not know its location nor exact position in the file, but are aware of its address or ID or some other attribute. In these instances, you can query the associated attributes to locate specific features.

The simplest means to query is by using the Find tool. The Find tool allows you to search all character fields of the active theme for the first occurrence of the specified string. Find is not case-sensitive, and can match on a partial string (e.g., St. Louis will match the search string "Louis"). If a match is found, the display is redrawn with the selected feature in the center of the display.

If the feature matched is not the desired feature, selecting Find again for the same string will find the next matching feature.

Except for the most specific and simple queries, selecting features by attribute is better served through the use of a logical query. A logical query can be used to select features based on how their columnar data matches with search criteria.

To select through a logical query, click on the Query Builder button from the button bar to call up the Query Builder dialog window. The Query Builder window displays a selection of fields, operators and values from which to construct a query. A query can be used to construct a new selected set, or to add to or remove from an existing set. When the query is executed, the selected elements are highlighted in the table and on the view.

The Query Builder dialog window.

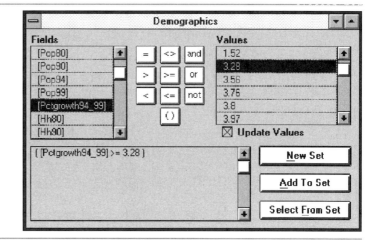

It is often easier to define features you do *not* want than those you do. The Switch Selection button from the Table button bar allows you to switch all features not selected to the selected set.

The Clear Selected Features icon is used to deselect the selected features in all active themes.

The Select All, Clear Selected Set, and Switch Selected Set icons.

The Undervalued GIS Skill of Database Management

As practitioners and promoters of a graphic system, we often focus and extol the virtues of visual analysis and the cartographic skills that advance the power of its usage. It is easy to overlook that many important GIS operations revolve not around graphics, but rather tabular data management.

Familiarity with database concepts and especially a background in SQL (Structured Query Language) can be a significant asset to the GIS operator. With that knowledge, the ease and power of conducting ad hoc logical queries increases immensely and enables much more rapid exploration of your overall spatial database. Essentially, strong tabular database knowledge is a perfect complement to strong cartographic knowledge.

Whether you are the manager of your shop or the newest kid on the block, we strongly encourage you to include database management training in your plans. If you operate with Access, be sure to train with Access. If you work with dBASE, train in dBASE. Many GIS training programs assume that you will pick up familiarity from your tabular database. Be sure to put SQL or other database management training on your schedule.

Locating the Selected Set

The results of a specific query from a table or theme may often not be readily apparent, even when the features are highlighted. The Zoom to Selected tool on the View button bar allows you to zoom to the extent of the selected features on the active themes.

The Zoom To Selected Features icon.

✔ *TIP: When executing a specific query against a theme containing a large number of features, zooming to the extent of the selected features before drawing may speed redraw time. In addition, opening the attribute table for the active theme will cause the number of selected elements to be displayed in the Table tool bar. This number can provide useful feedback on the outcome of a query before the elements are examined or displayed.*

Logical Queries on Themes

In the examples above, all features present in the original theme were available for query. In many instances, however, restricting a theme to a subset of the available features is desirable. For example, a data set of soils information might be used to extract a theme containing only soils with a high water table.

To define a logical query to extract features from a theme, make the theme active, and select Properties from the Theme pull-down menu to access the Theme Properties dialog window. Click on the Definition icon, and click on the Query Builder icon to call up the logical query dialog window.

A logical query is constructed in the same manner as when using the Query Builder to select features from an active theme, with one difference. As this query defines the base set of features comprising the theme, there is no option to add to or remove from the selected set.

When the query is complete, press OK to accept the query. The query will then appear in the Theme Properties definition box. Press OK in the Theme Properties dialog window to apply this query to the theme. Only those features will now be available in the theme for display and query.

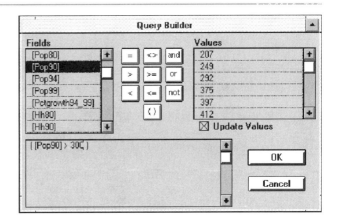

The Theme Query Builder dialog window.

A Review of Logical Queries

Selecting features using the Query Builder has been mentioned many times. For those not familiar with logical queries, some elaboration is in order.

All logical queries are comprised of *fields*, *operators*, and *values*. Complex queries can be built from these elements, but all queries share this basic form. At its most basic, a logical query is comprised of one field, one operator, and one value. For example, the query

    (block_group = 3194.102)

can be used to select any features or records for which the *block_group* is 3194.102.

The query

    (acres > 640)

can be used to select all features with an area greater than 640 acres.

Complex queries can be constructed by combining simple expressions with the And or Or operators. For example, the query

    (acres > 640 and owner = "Jones")

can be used to select all parcels greater than 640 acres owned by Jones. Note that character fields must be enclosed in double quotes.

The following operators are available in the Query Builder:

    = (equals)
    > (greater than)
    < (less than)
    <> (not equal to)
    >= (greater than or equal to)
    <= (less than or equal to)

For complex queries, the following joining operators are available:

    and (both expressions are true)
    or (either expression is true)
    not (excluding the following expression)

The following wildcard operators are available:

    * (multiple character wildcard)
    ? (single character wildcard)

Next, the following mathematical operators are available:

```
+ (addition)
- (subtraction)
* (multiplication)
/ (division)
```

Note that all logical expressions evaluate from left to right *regardless* of the mathematical operator. Thus, the expression

```
( 3 + 8 * 7 )
```

will yield the value 77, not 59, as might be expected.

Parentheses are used to force an expression to be evaluated first. Thus, the expression

```
( 3 + ( 8 * 7 ) )
```

will yield the value 59.

Queries can be used to compare the values of two fields. For example, the query

```
( [value80] < [value90] )
```

will return the records for all parcels which have decreased in value from 1980 to 1990.

Calculations can be performed on fields as well. For example, the query

```
( [value90] / [acres] > 10000 )
```

can be used to return the records for all parcels valued at greater than $10,000 per acre.

> NOTE: In ArcView (`[name] = "*Main Street Cafe*"`) will match "Main Street Cafe" in your table, regardless of how many blanks the field is padded with. If you are in doubt as to whether a particular field contains padding blanks, the list of values produced by selecting the field in ArcView Query Builder will contain all blanks from that field. Double-clicking on a value to add it to the query will cause the requisite number of blanks to be included within the quoted string.

> Our best advice to new users is to dive into the Query Builder and experiment. Once the structure and vernacular of queries are understood, the flow becomes easy. Do not worry if seemingly simple queries occasionally fail to fire off as expected. With some practice, you will soon be executing them with ease.

Exercise 5: Making a Smarter Map

By the end of this exercise, you will have produced visually appealing and powerful maps!

The first step is to open the *ch6.apr* project file in your working directory. We are not asking you to open the project you saved at the end of the previous exercise, because we added work to the file. As you look at the current version of the project, you will notice three changes: the two IHOP survey themes were combined into a single theme, an Arterial Traffic theme containing traffic count information was added, and a MicroVision theme containing Equifax MicroVision Segment information was added. In addition, an initial classification on the IHOP and Traffic themes has been performed.

As we continue with the market research project by combining the IHOP surveys into a single theme, we now have a single point theme locating customer responses classified by amount spent. The next step is to associate these responses to the block group in which they are located. From there, we can compare these responses to other data we have mapped by block group, such as demographic data.

To associate the survey response points to the block group, you need to join the *IHOP* point theme to the *Blkgrp* theme. This spatial association requires a special type of join, known as a *spatial join*.

1. The first step in the spatial join is to open the attribute tables for the *IHOP* and *Blkgrp* themes. Next, highlight the Shape field of each table. Because we wish to associate a block group with every *IHOP* survey point, the *Blkgrp* attribute table will be the source table, and the *IHOP* attribute table the destination table.

Exercise 5: Making a Smarter Map 125

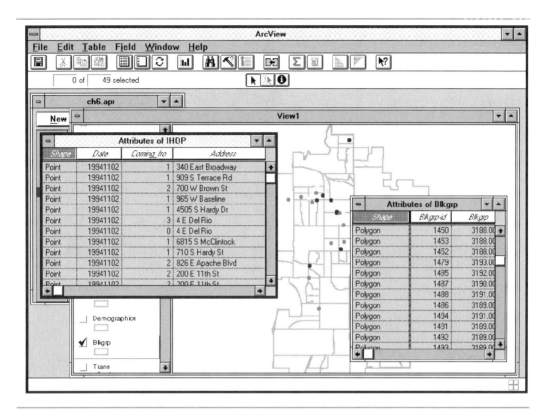

The IHOP and Blkgrp theme attribute tables ready for the join.

2. Select Join from the Table pull-down menu. Upon completion, the *Blkgrp* attribute is associated with every response point in the *IHOP* theme.

At this point, we wish to summarize the amount spent per respondent for each block group.

1. With the *Attributes of IHOP* table open, highlight the Blkgrp field name. Note that when you do so, the table manipulation tools—Sort Ascending, Sort Descending, and Summarize—are no longer grayed out.

Chapter 6: Data Queries

Attributes of IHOP					
Av_add	Av_status	Av_score	Av_side	Blkgrp-id	Blkgrp
6522 S FARMER	M	83	R	1751	3200.015
625 W 1ST ST	M	99	L	1447	3188.004
829 W BROWN	M	83	L	1447	3188.004
1239 E BROADMOR	M	83	R	1579	3195.001
155 W 3RD ST	M	99	L	1420	3187.002
222 E APACHE BLV	M	99	L	1536	3190.002
1242 E CONCORDA	M	83	L	1579	3195.001
295 E APACHE	M	83	R	1487	3190.001
919 E LEMON ST	M	99	R	1498	3191.003
401 E APACHE	M	83	R	1536	3190.002
1215 S FOREST	M	83	L	1487	3190.001
625 W 1ST ST	M	99	L	1447	3188.004
330 E ORANGE	M	83	L	1487	3190.001

The results of the spatial join to the IHOP theme.

2. Click on the Summarize tool. From the field list, select the Spent field as the field to summarize, and Average as the option to summarize by. Click on Add to add this selection to the summary table. Specify the name and path for the output table, and then click OK to proceed.

Upon completion, the output table, *spent.dbf*, is created. The table contains three fields: *Blkgrp*, *Count*, and *Ave_spent*.

✖ *WARNING: Due to the manner in which a saved project is reopened, accessing a theme resulting from a spatial join in a restored project can cause a fatal error. To prevent the error, it is necessary to remove all spatial joins from the project before saving.*

Exercise 5: Making a Smarter Map 127

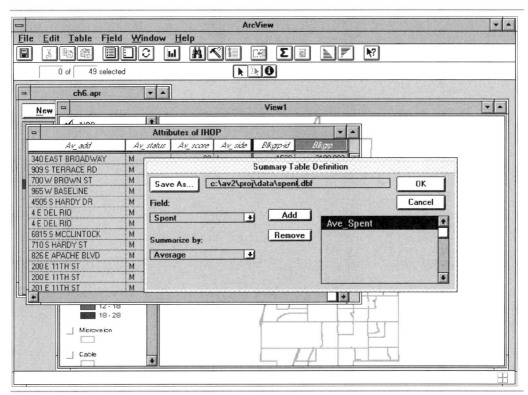

The Summary Table Definition dialog window.

Because we have successfully created the summary table, we can remove the spatial joins now.

1. Click on the title bar of the *Attributes of IHOP* table to make it active.

2. From the Table pull-down menu, select Remove All Joins. The Blkgrp field is no longer associated with the *Attributes of IHOP* table.

3. Close the *Attributes of IHOP* table and return to the *spent.dbf* table.

128   Chapter 6: Data Queries

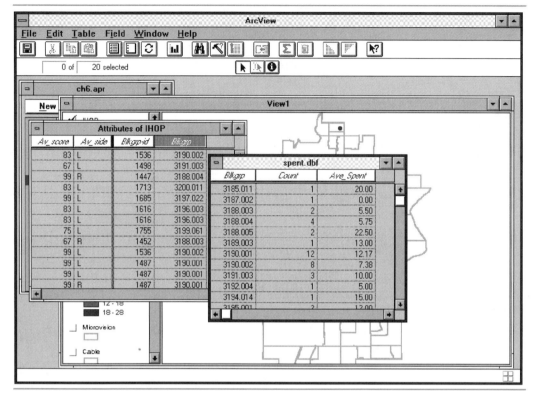

The resultant spent.dbf summary table.

You now have a table containing the average amount spent by block group that you can relate back to the *Blkgrp* theme.

1. Make the *Blkgrp* theme active, and copy the theme using Copy Theme and Paste from the Theme pull-down menu.
2. The new theme appears at the top of the Table of Contents. Click on this theme to make it active, access the Theme Properties dialog window, and change the name of the theme to *Avg Amount Spent*.
3. Open the *Attributes of Avg Amount Spent* and *Spent.dbf* tables, and Join *Spent.dbf* to *Attributes of Avg Amount Spent* using the common *Blkgrp* field.

Exercise 5: Making a Smarter Map

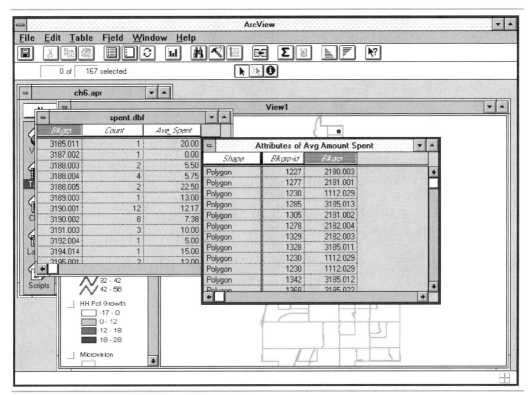

The spent.dbf and Attributes of Avg Amount Spent tables ready to join.

Once the tables are joined, you can classify the *Ave Amount Spent* theme on *Avg_spent*.

1. Double-click on the theme to bring up the Legend Editor.

2. Classify the theme in four classes using quantile classification, and Apply. (The incremental project has been saved as *ch6a.apr*.)

130   Chapter 6: Data Queries

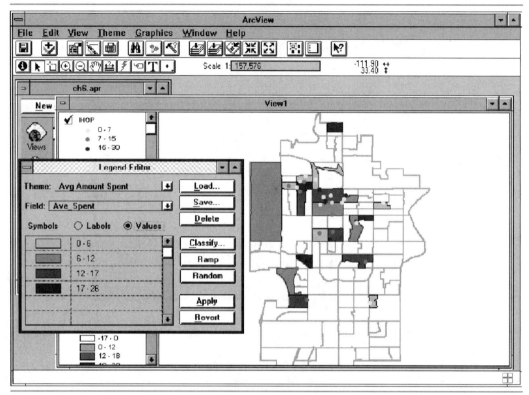

The applied classification, showing average amount spent along with the IHOP survey response points.

At this juncture, you have successfully generated an indication of how the amount spent varies by neighborhood. For the next series of analyses, we wish to examine the area immediately adjacent to the client's location. First, you need to determine the location of the client by querying the *Restaurants* point theme.

1. Turn on the *Restaurants* theme, and click on the theme to make it active.
2. To access the Query Builder, click on the Query Builder icon from the button bar, or select Query from the Theme pull-down menu.
3. Double-click on the *Name* field. The Values list at the right will be populated with the values for *Name* from the *Restaurant* theme attribute table.

4. Click on the equals (=) operator, and then scroll down the list until you locate the name *INTERNATIONAL HOUSE OF PANCAKES*. Double-click on this entry.

5. When the query statement is complete, click on the New Set button. The location of the client, the International House of Pancakes, will be highlighted. Make sure that the *IHOP* theme is turned off so that you can see the result of this query.

The Query Builder for the Restaurants theme.

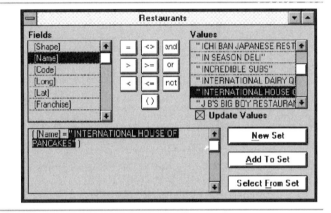

6. To select all demographics and customers within a three-mile radius of the client's location, click on the Graphics tool (on the far right of the tool bar), and select the Circle tool.

7. Click on the IHOP location, and drag a circle until the status bar reports a radius of 3.00.

• *NOTE: To measure the radius distance in miles, the Distance Units for the View must first be set to miles. If the units have not been set to miles, or you are not sure whether units have been set, access the View Properties dialog window from the View pull-down menu, and select Miles from the list for Distance Units.*

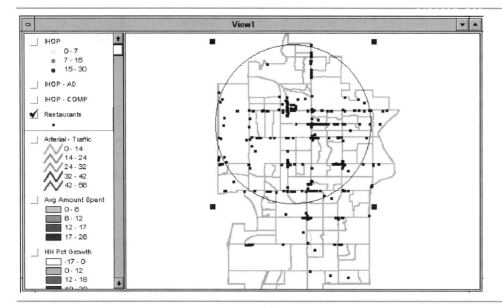

The three-mile radius circle centered on the IHOP location.

We are now ready to use this circle to select block groups falling within the three-mile radius. But look closely: something is amiss with the circle.

The arterial streets in Tempe are located on a mile grid. The IHOP restaurant is located almost exactly halfway between the mile grid arterials in both the north/south and east/west directions. The circle describing the three-mile radius should also fall between the mile arterials. Note that although this is the case to the north and south, to the east and west the circle falls directly on the mile arterial, which is a distance of only 2.5 miles.

This distortion results from the fact that we are still working in geographic coordinates. Although we have set our map units to miles, geographic coordinates do not constitute a true projection, and distances measured are not uniform in both the x and y directions. Thus, while working in geographic coordinates is adequate when *relative* positions are sufficient, when accurate distance and area calculations are desired, it is necessary to switch to a projection where accurate area and distance is maintained. For our project area, Arizona State Plane coordinates work.

From the View Properties window, select Projection to access the Projection Properties window. Under category, select State Plane - 1983, and for Type, select Arizona, Central. Click OK to apply.

Exercise 5: Making a Smarter Map

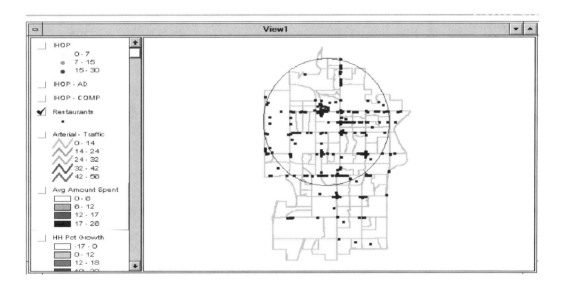

The Projection Properties dialog window.

View1 projected to Arizona State Plane coordinates.

Note how the circle is now quite clearly an oval. We will need to re-do the three-mile radius. Select and delete the old circle, and add a new circle with the same three-mile radius.

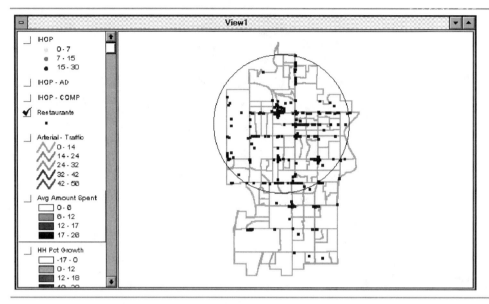

The new three-mile radius circle for View1.

We can now identify the block groups within the three-mile radius which spent the greatest amount per capita. First, make the *Avg Amount Spent* theme active. Next, use the selection tool (the arrow) to select the circle. At this point, we can use the Select Features by Selected Graphics tool.

The Select Features by Selected Graphics tool.

Based on our initial classification, we had determined that $15 is a suitable cut-off point for identifying block groups with a high average amount spent. Access the Query Builder, and construct the query *Ave_spent >= 15*. Choose the Select from Set to extract this new set from the currently selected block groups. Six block groups are now selected.

Exercise 5: Making a Smarter Map 135

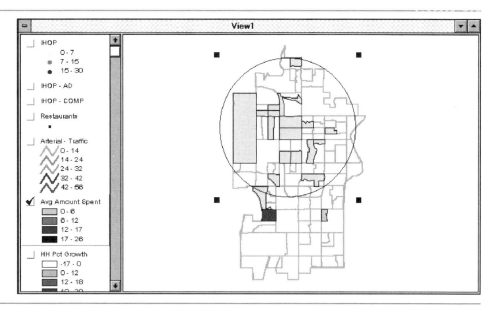

The selected block groups from the Avg Amount Spent theme.

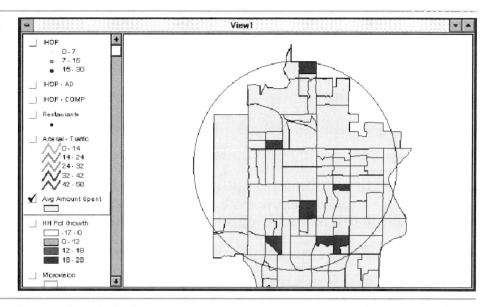

The six selected block groups with average amount spent at least $15, and located within a distance of three miles.

We have displayed the polygons in a uniform color by removing the classification on the *Avg Amount Spent* theme, selecting <none> for the Field name in the Legend Editor. We have assigned a light shade to the theme, and subsequently changed the color used to draw selected features to a dark color. We accomplished this by accessing the Project Properties dialog window from the Project pull-down menu in the Project window, and setting the color for display of selected features to a dark color using the Specify Color sliders. (The incremental project has been saved as *ch6b.apr*.)

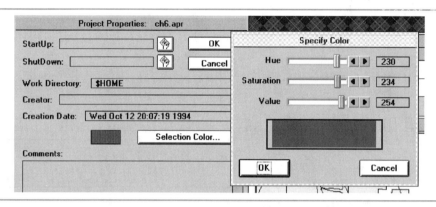

The Specify Color sliders from the Project Properties dialog window.

Having identified six favorable block groups from our customer base, we can now learn more about the demographics and lifestyles of some of our best customers. To accomplish this, we will access the MicroVision segmentation data.

MicroVision is a micro-geographic targeting system that uses aggregated consumer demand and census data to classify every household in the United States into one of 50 unique market segments. Each market segment consists of households at similar points in the life cycle that share common interests, purchasing patterns, financial behaviors, and needs for products and services.

Exercise 5: Making a Smarter Map   137

To identify the corresponding block groups from the MicroVision theme for the six selected block groups of the *Avg Amount Spent* theme, we make use of the ability to select features from a second theme based on the location of selected features from the first theme (i.e., Select by Theme).

1. First, make the MicroVision theme active.
2. From the Theme pull-down menu, choose Select by Theme.
3. Choose to select features that Are Completely Within the selected features of *Avg Amount Spent*, and then click on New Set.

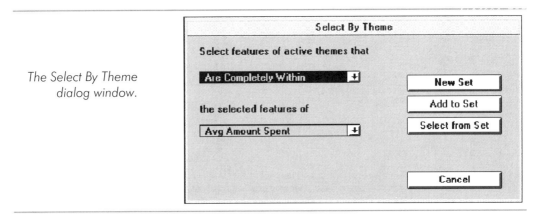

The Select By Theme dialog window.

4. Turn off the *Avg Amount Spent* theme, and turn on the MicroVision theme to display the results of this selection.

138   Chapter 6: Data Queries

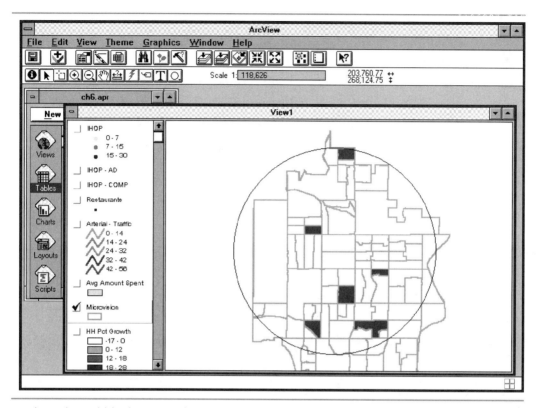

The selected block groups from the MicroVision theme resulting from the theme on theme selection.

While we could use the Identify tool to determine the primary MicroVision segment associated with each block group, here we elect to label each block group with the associated value using the Auto-Label tool.

The Auto-Label tool labels selected features of the active theme using attributes from the designated field. The field to use for labeling, as well as the placement of the text label, is controlled from the Theme Properties dialog window. To access these properties, and use the Auto-Label tool, take the following steps:

1. Activate the Theme Properties dialog window, and select the Text Labels icon to access the label properties.
2. From the Label Field list, select the *Prim_mv* field.

The Text Label properties from the Theme Properties dialog window.

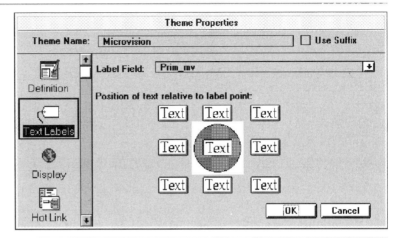

3. From the Theme pull-down menu, select Auto-Label. Text labels corresponding to the values for *Prim_mv* are placed in each of the six selected block groups. These text labels are graphics elements, and therefore can be repositioned for greater legibility.

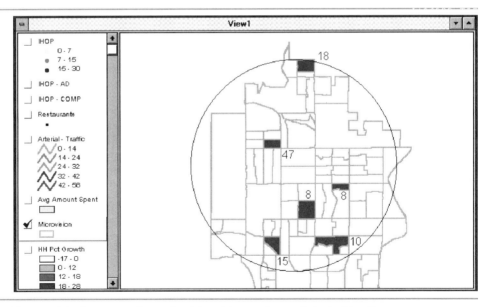

The six block groups from the MicroVision theme labeled by Prim_mv.

We can now refer to the demographic tables for additional information on each of the identified MicroVision segments. (For details on segments, see Appendix C, "About the MicroVision Segments.")

Now we are going to identify heavy traffic arterials adjacent to these block groups as candidate areas for outdoor advertising. Upon examining the initial classification, we determine that a threshold of 40 (40,000 vehicles per day) is a good cut-off point. To enhance the display of the classification, take the following steps:

1. Double-click on the *Arterial Traffic* theme to access the Legend Editor.

2. Reclassify the theme to two classes, and adjust the class ranges to 0 - 40 and 40 - 56. To improve the display, we have used a light gray line for the lower class, and a heavy black line for the higher class.

3. To highlight the traffic data, we have changed the symbol for the MicroVision theme so that polygon outlines are not drawn. This is done by selecting <none> for line thickness from the Shade Palette. Thus, only the selected block groups from the MicroVision theme will draw. (The incremental project has been saved as *ch6c.apr*.)

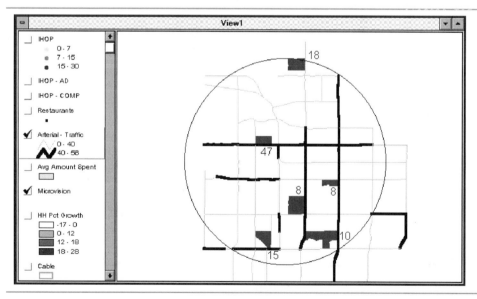

The classified Arterial Traffic theme displayed with the selected block groups from the MicroVision theme.

Exercise 5: Making a Smarter Map   141

Our last step is to use the *Restaurants* point theme. Rather than display the location of all restaurants, we can make use of an additional field contained in this table which identifies the type of business. This code, known as the SIC (Standard Industrial Classification), is stored in the Code field of the *Restaurant* attribute table (see endnote). By querying the restaurant theme table, we determine that the SIC for IHOP is 581260. It is logical that examining only those restaurants with an SIC of 581260 would give us a better picture of competitors' locations.

To limit our display to family restaurants, we will apply a logical query on the *Restaurant* theme.

1. Click on the *Restaurant* theme in the Table of Contents to make it active.
2. From the Theme Properties dialog window, click on the Definition icon. Select the Query Builder tool icon.
3. Construct the query *Code = 581260* (the SIC code for family restaurants). Fifteen features of a total 388 are selected.
4. The *Attributes of Restaurants* table reveals that the selected set of restaurants includes Denny's, JB's, Howard Johnson's, Village Inn, and Shoney's, all of which were previously identified as IHOP competitors. As before, you can adjust the symbol for greater clarity.

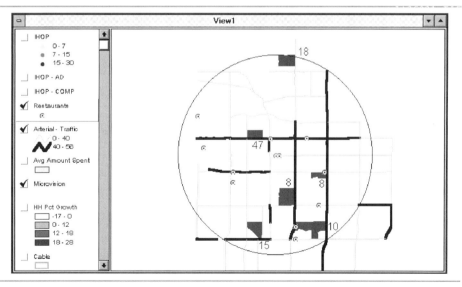

The previous display with the location of competing restaurants added.

The map in the illustration above is usable, but enhancements could improve its visual presentation. For example, we could replace the MicroVision segment number with a brief descriptor, and label some of the competing restaurants.

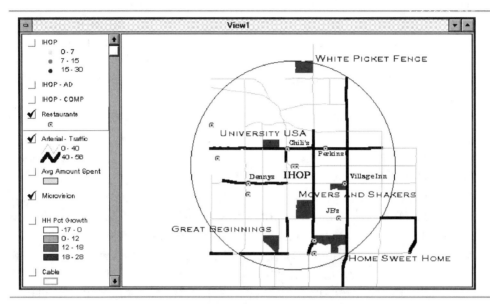

The same map with enhancements.

There is more we could do here. For instance, we have not yet explored the cable theme, which could yield valuable information on the cable channels most useful for reaching our targeted market through advertising. But we think we have proven our point that spatial analysis and mapping tools can greatly enhance traditional market research.

Endnote: This information was derived from the Equifax National Decision Systems' Restaurant-Facts™ database. Certain components of the Business-Facts® database are derived using data obtained from American Business Information Inc., Omaha, Nebraska, © 1994, and Disclosure Incorporated, Bethesda, Maryland, ©1994. The primary business data source for the Business-Facts™ database is licensed from American Business Information Inc., Omaha, Nebraska.

Chapter 7

Charts

When working with thematic data, the available options for thematic query and classification are often not sufficient to fully reveal the relationships inherent in the data. The charting capabilities in ArcView provide additional tools for representing the attributes on your map.

One of the strengths of charts in ArcView is that they are *dynamic*. The chart represents the current data in the table. If the table is updated, the chart will reflect the change.

A chart can represent a selected set of records in a table as well as the entire table. As the selected set is changed, the chart immediately reflects this change. Its dynamic nature allows a chart to function as an especially powerful visualization tool during interactive query.

Six types of charts are available: area, bar, column, line, pie, and xy scatter diagrams. When a table and records have been selected for charting, the user can toggle between chart types to determine which is most effective in representing the data.

Creating a Chart

There are two basic approaches to creating a chart. The first approach, which might be considered more direct, is to click on the Chart icon from the Project window to make charts the active project component, click on the New button and then select the table you wish to work with from the list box containing chartable tables.

However, if you wish to chart from a table subset, such as the results from a logical or spatial query, you can approach chart creation in another

144 Chapter 7: Charts

fashion. In these instances, with your table already active and a selected set in place, you could simply select Chart from the Table pull-down menu.

Both approaches will activate the Chart Properties dialog window, where you are presented with additional choices on chart design. From the Chart Properties dialog window, the field or fields to chart are selected, and optionally, the field to be used to label each series or group. When the desired fields have been selected, clicking OK will create the chart.

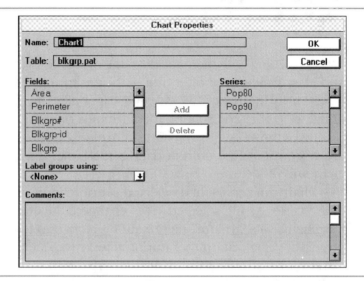

The Chart Properties dialog window.

By default, a column chart is created with the chart's *data series* formed from the selected records in the table.

Clarification of Data Markers, Series, and Groups

In discussing the creation of charts, data markers, data series, and data groups are frequently mentioned. It is easy to get confused, even with the aid of the on-line help and the ArcView manual. While it is possible to create a perfectly usable chart without a full comprehension of these terms, given the frequency with which they appear in the documentation we feel it is worth the effort to come to grips with them.

A *data marker* is a feature on a chart—column, bar, area, pie slice, or point symbol—which represents the value of a specific field for a specific record in a table. It is analogous to a cell in a spreadsheet. The data marker can represent the actual value for this cell, or may be expressed as a percentage or logarithm.

To best understand *data series* and *data groups*, we need to examine an illustration. In our example, we selected 10 records from a table, and intend to chart two fields from these records. This gives us 20 data markers for the chart. The records represent neighborhoods, and the two fields are 1980 and 1990 population counts.

In the chart at the left, the data series was formed from the table's records. Each instance, or data marker, is represented by a column. The data group in this example is the field. We are charting two fields: *Pop80* and *Pop90*. Thus, the chart contains two data groups, each containing 10 instances. In the data series legend, each record in both data groups is assigned a different color.

In the chart at the right, the data series has been formed from the fields from the table. Again, each instance, or data marker, in the series is represented by a column. The data group in this example is the record. Because we are charting 10 records, the chart contains 10 data groups, each containing two instances corresponding to the two fields, *Pop80* and *Pop90*. In the legend, each field in the 10 data groups is assigned a different color.

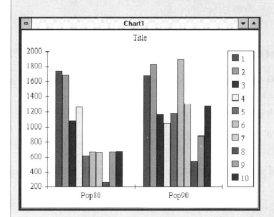
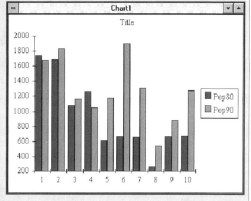

Two charts created from the same selected records. The data series on the left was formed from records, and the data series on the right was formed from fields.

The chart data series, then, is the comparative unit in the chart. If the data series is formed from records, it contains the full set of selected records for

> a specific field. If the data series is formed from fields, it contains the full set of selected fields for a specific record. The chart data group is the aggregating unit in the chart.

➥ **NOTE:** *There are several caveats to bear in mind as you begin charting. For example, only numeric fields are available for charting. Next, certain types of charts have limitations on the number of rows they can work with. If you receive a message such as "CHART: There is not enough space to plot the chart; check the format parameters and/or resize the chart," switch to another chart format or make your window larger. ArcView does not allow you to scroll around a chart that overflows the screen, so be prepared to alter your format to one that can be adequately displayed all at once.*

Making Changes to a Chart

There are three categories of changes that can be made to a chart: changing the selected set of records displayed in the chart, changing the chart type, and changing the way the chart elements are displayed.

Changing the Selected Set

In ArcView, a *data marker* is created on a chart for every selected record in the charted table. Changing the number of records selected from the table will accordingly change the resultant chart. You can change your selection interactively by clicking on records in the table, by logical query on the table via the Query Builder, or by spatial selection when charting on a theme attribute table. Regardless of the technique you use, the result will be the same: the chart will "forget" references to your old selections and depict only those records that match your latest query criteria.

Changing the Chart Type

The six chart types available in ArcView differ in the way they represent thematic data. Depending on your data, you may find one or two chart types to be superior for illustrating what you wish to portray. For example, bar charts are useful in comparing relative amount or change, such as population growth, while pie charts are useful in illustrating composition, such as the ethnic makeup of a district.

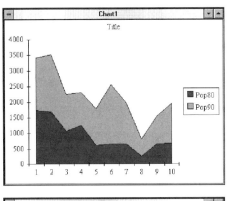
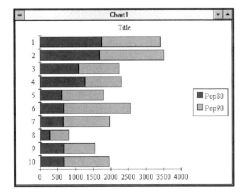
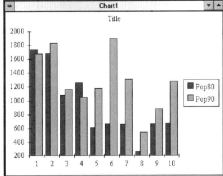
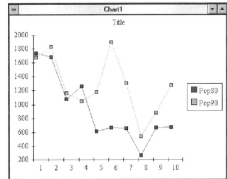
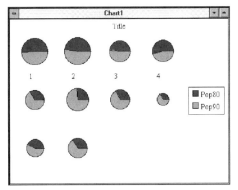
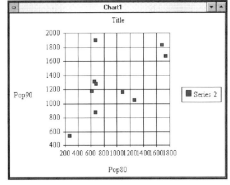

The same data displayed using six different chart types: area chart (top left); bar chart (top right), column chart (center left), line chart (center right), pie chart (bottom left), and XY scatter chart (bottom right).

Changing the Display of Chart Elements

Once the chart type and the desired records from the table have been selected, you can also change how the individual chart elements will appear. Default styles are seen within the chart types. These choices include how to display individual data markers, whether to display grid lines, and whether to use a linear or logarithmic scale. Chart elements, including the title, legend, and axis properties, can also be edited. Finally, the Color Palette can be used to control the display of chart elements and data markers.

Using Charts Interactively

One of the strengths of charts in ArcView is that they are dynamic. As the underlying data changes or selected sets change, values in the charts will change. With this capability, the user can investigate data using the charts for immediate display of the attributes associated with currently selected features.

Locating Theme Elements with the Identify Tool

In a manner similar to identifying map elements, the Identify tool allows you to click on a data marker on a chart, locate the element in a table or on a view, and examine the attributes associated with the data marker. Clicking on a data marker in a chart with the Identify tool active causes the corresponding record in the table to flash. The Identify window is displayed, showing all the attribute data for the feature. If the chart is built on a theme attribute table that is currently displayed, the feature will blink on the map as well.

Using Charts with the Select Tool

Because a chart displays the attributes of the selected features, it can become a powerful tool for visualizing the results of a spatial query, particularly when a chart displays more than one attribute. In the following example, a chart has been constructed on the attribute table for a theme displaying demographic data. From this table, the fields representing average household income have been selected to be charted using a pie chart.

With the chart displayed alongside the view window, the Select tool is used to select specific census block groups from the demographic theme. As the feature is selected, the chart displays the proportional distribution of household income for each class. The illustration on the left shows the results for an area near Arizona State University, and on the right, a suburban neighborhood in south Tempe. Depending on the type of chart selected, the results of queries on individual features or groups of features can be displayed.

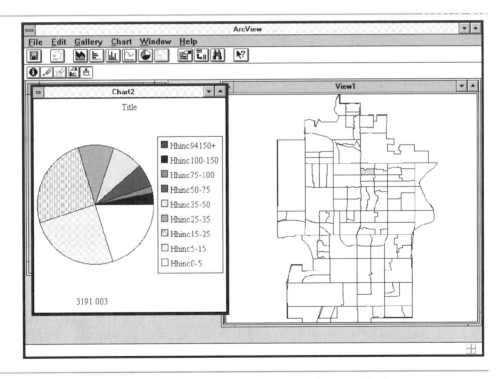

Pie charts showing household income distribution for two block groups. A block group adjacent to the university campus, dominated by apartments catering to students appears in the illustration above. A block group in McCormick Ranch, an upscale residential community, appears in the illustration below. The corresponding block groups are highlighted in the view at the right.

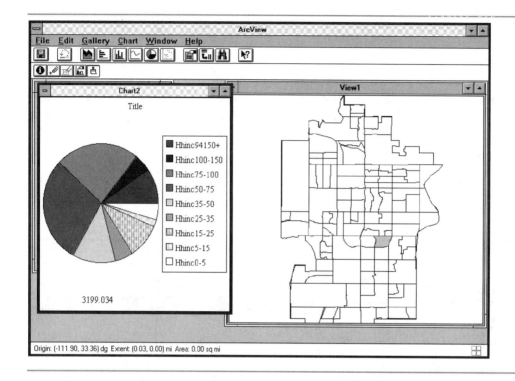

More on Chart Characteristics

We have seen the strengths of charts in ArcView, that is, their dynamic properties and the ability to use them in interactive queries. As mentioned earlier, however, these strengths are not without limitations.

The primary limitation is that charts are tied to specific records in a specific table. While charts are very effective for depicting specific values, they do not handle ranges well. In brief, while you can easily classify themes on attribute values and display the results of these classifications, attributes grouped on these thematic classes are charted with difficulty.

For example, assume that you want to chart the average household income associated with each of five classes resultant from a classification on population growth. You must first add an item to the theme's attribute table. Next, you have to manually select each set of features from the theme corresponding to the population growth classes, code each record in each set with a class identifier, and then use ArcView's statistical capabilities to calculate the average household income for each population growth class and store the results in a new table. The new table, which contains five records with fields corresponding to the population growth class and average household income, is the table that is finally charted to produce the results. Although this operation is not difficult, it is also not one which can be executed repeatedly without careful concentration.

If you have a need for performing the above operation on a regular basis, customization through ArcView's Avenue scripting language is recommended. An Avenue script can be used to automate an otherwise cumbersome task and integrate it into a larger ArcView project. Selected Avenue features are examined in Chapter 11.

Exercise 6: Working With Charts

After the exercise in Chapter 6, you may have thought we had taken our project about as far as it could go. In reality, much territory remains to be explored, and some of it is ideally suited for ArcView's charting capabilities.

1. Begin by opening the *CH7.apr* project file in your working directory.
2. Turn off all themes, and then turn on the MicroVision theme. Click on the MicroVision theme to make it active.
3. Copy the MicroVision theme using the Copy Theme and Paste selections from the Edit pull-down menu. Click on the copy to make it the active theme.
4. Access Theme Properties, and change the name of the new theme to *Mv_spent*.
5. Note that the theme still has the six block groups selected, as well as the text labels identifying the MicroVision segments and competitor restaurants. The block groups and text labels were copied along with the theme. To delete these elements, select them using the Graphics Select tool (the arrow), and then choose Delete Graphics from the Edit menu.
6. Click on the Unselect tool from the button bar to remove all features from the selected set.

7. Although the *Mv_spent* theme is still turned on, the only thing visible is the three-mile radius circle. This is because the shade color for the *Mv_spent* theme is set to *transparent*, and the outline line width is set to *<none>*. The result of the two properties settings is to make the theme invisible. Double-click on the theme to access the Legend Editor, and change the outline width back to *1*. After you click on Apply, the block group outlines should draw.

To further investigate the restaurant customers, our first objective will be to associate the MicroVision segments with the amount spent. In the previous exercise, we produced a summary table calculating the average amount spent by block group, or *spent.dbf*. By joining the summary table to the *Mv_spent theme*, we can examine how much was spent on average by each MicroVision segments.

1. Open the attribute table for the *Mv_spent* theme.
2. Click on Tables in the Project window. From the list of available tables, select and open *spent.dbf*.
3. Highlight the *Blkgrp* field on both tables. Join the *spent.dbf* table to the *Attributes of Mv_spent* table.
4. Select the Query tool. Using the Query Builder, construct the query *Count > 0*, and click on New Set. This action will select the block groups for which we had survey responses. The status line shows that 20 of 167 features are selected. The selected features are highlighted on the view as well as in the *Mv_spent* attribute table.

Instead of classifying the data as we would do in a map, let's generate a chart to examine the relationship between the MicroVision segment and amount spent.

Exercise 6: Working With Charts 153

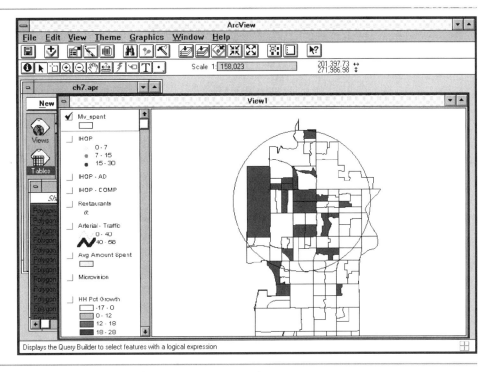

The selected features from the Mv_spent theme.

1. From the Project window, select the Chart icon, and then click on New.
2. A dialog box prompts you to Pick a Table. From the pull-down list, select *Attributes of Mv_spent*, and click OK.
3. You have accessed the Chart Properties dialog window. From the scrolling list of Fields, select *Prim_mv* and *Ave_spent*. Click on Add to add each to the list of Groups.
4. From the Label Series Using list, select *Prim_mv*. For the name, use *MV Spent*. Click on OK to accept these choices.

154   Chapter 7: Charts

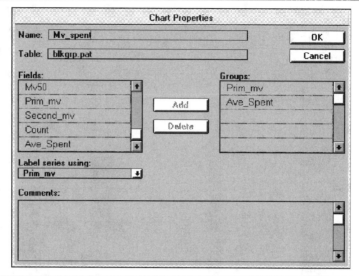

The Chart Properties dialog window.

You are probably wondering why we are working with two variables when we only want to plot *Ave_spent*. The reason is that when charting from only one field, ArcView will not allow you to label the bar lines from one field with the values from another. There is a work-around, but it necessitates starting with two fields.

By default, you are presented with a bar chart, with the data series formed from the selected records. Two data groups are present which correspond to the charted fields, *Prim_mv* and *Ave_spent*.

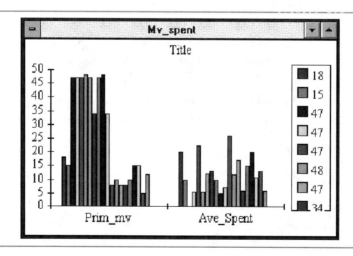

The default chart using the Chart Properties choices entered above.

Exercise 6: Working With Charts 155

The above chart is not very effective. We can improve the chart by switching the series being formed from *records* to *fields*.

1. Click on the Series From Records / Series From Fields icon in the button bar.

The Series From Records / Series From Fields icon.

2. Now we have a glimpse of MicroVision expenditures. Resize the chart window as needed to make the chart more legible.

The new chart with the data series formed from fields.

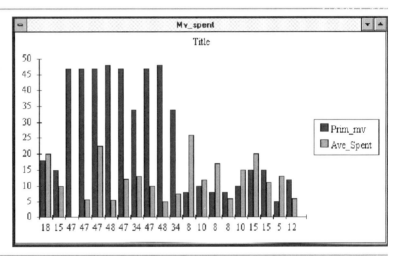

3. From the Chart menu, select Properties to return to the Chart Properties window.
4. Click on *Prim_mv* from the list of Groups (Series), and select Delete. Click OK to apply this change. The chart series is still formed from fields, but now only one field, *Ave_spent*, is displayed. The data markers, however, are still labeled with the values for *Prim_mv*, which is just what we want. This is the outcome of the work-around we mentioned earlier. (The incremental project has been saved as *ch7a.apr*.)

The revised chart with one data series formed from the Ave_spent field.

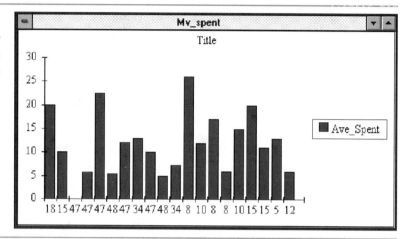

Upon taking a closer look at the above chart, we can identify several records with the same value for *Prim_mv*. What we would *really* like to see is a chart showing average amount spent by *Prim_mv*. To accomplish this, we need to further manipulate the data.

1. Close the chart, and return to the view.

2. We need a new summary table by block group, this time not *averaging* the amount spent, but *summing* it. To produce this summary table, we first need to recreate the spatial join we used in Chapter 6 (Exercise 5). Open the attribute tables for the *IHOP* and *Blkgrp* themes. Highlight the Shape field in each table. Click on the title bar of the *Attributes of IHOP* table to make it the active, or destination table, and then select Join from the Table pull-down menu. As before, the *Blkgrp* attribute is now associated with every response point in the *IHOP* theme.

3. Click on the Summary tool to create a summary table.

4. For Field, select Spent, and for Summarize By, select Sum.

5. Click on Add. The entry *Sum_Spent* is added to the summary field list. Enter the output file information, or *sumspent.dbf*, and click OK. (Be sure to verify that the *sumspent.dbf* table is added to your working directory.)

• **NOTE**: As before, we need to remove the spatial join before saving the project. Click on the title bar of the Attributes of IHOP table to

Exercise 6: Working With Charts 157

make it active, and select Remove All Joins from the Table pull-down menu.

6. Close the *Attributes of IHOP* table, and open the attribute table for the *Mv_spent* theme.
7. Use the *Blkgrp* field to join the *sumspent.dbf* table to the *Attributes of Mv_spent* table.
8. We are now ready to create the new summary table. If you no longer have the 20 records selected previously from the *Attributes of Mv_spent* table, re-select them using the query mentioned above.
9. Highlight the *Prim_mv* field in the *Attributes of Mv_spent* table, and click on the Summary tool.
10. For Field, select *Sum_spent* and Count. For both, select Summarize by Sum. Add both to the summary field list, and save this new table with the name *mvsumsp.dbf*. Click OK to create the table.

[handwritten margin note: Add one & then the other (one at a time)]

The resultant table should contain nine records, each associating a *Prim_mv* class with the total amount spent, as well as reporting the total count of survey responses for the *Prim_mv* class. However, what we really want to know is the per capita amount spent, or the total amount spent divided by the total count. We have these fields in our table, but we need to add a new field to contain the results of the division.

By default, ArcView prohibits table edits. However, we can add a field to a table by first making the table *editable*.

1. From the Table pull-down menu, select Start Editing. Note that the column headings shift from italic to regular font style.
2. From the Edit pull-down menu, select Add Field. The Field Definition window will appear.
3. Enter *Pcap_spent* for the field name, select Number for Type, set the width at *8*, and the decimal places at *2*. Click OK to add the field to the table.

158   Chapter 7: Charts

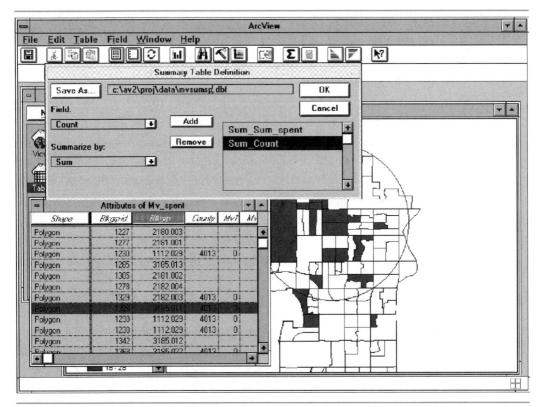

The Summary Table Definition to create the mvsumsp.dbf table.

The Field Definition window.

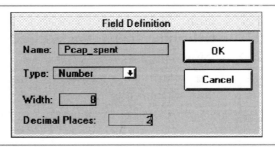

Exercise 6: Working With Charts   159

4. To calculate the new values for *Pcap_spent*, highlight the field name, and then click on the Calculate tool from the button bar (the tool that resembles a calculator).

5. The Field Calculator dialog window looks much like the Query Builder window. Form the expression *Pcap_ spent* = *Sum_Sum_ Spent / Sum_Count* by double-clicking on the appropriate Fields and Requests. When you have properly formed the expression, click OK. The results of the calculation are immediately written to the target field.

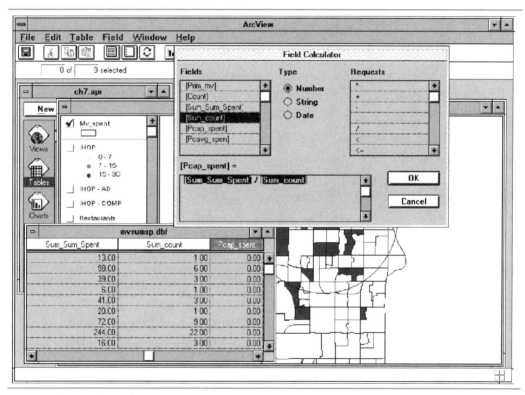

The Field Calculation dialog window.

6. Examine the table to ensure that your results are as anticipated.

Results of the above calculation.

Prim_mv	Count	Sum_Sum_Spent	Sum_Count	Pcap_spent
5	1	13.00	1.00	13.00
8	3	98.00	6.00	16.33
10	2	39.00	3.00	13.00
12	1	6.00	1.00	6.00
15	3	41.00	3.00	13.67
18	1	20.00	1.00	20.00
34	2	72.00	9.00	8.00
47	5	244.00	22.00	11.09
48	2	16.00	3.00	5.33

7. Select Stop Editing from the Table pull-down menu to lock the table to further editing. When you do so, the field names for the table will return to the italicized font style.

We now have the field we want to chart. To create a new chart, take the following steps:

1. From the Project menu click on the Chart icon, and then New. Select the *mvsumsp.dbf* table.
2. For Fields, select *Prim_mv* and *Pcap_spent*, and for Label Series Using, select *Prim_mv*.
3. Give the chart a meaningful name, such as *Per Capita Spent by MicroVision Segment*. Click OK to apply.
4. Switch the series from records to fields. (called series)
5. Access the Chart Properties window again, delete the group *Prim_mv* from the chart, and apply. (The incremental project has been saved as *ch7b.apr*.)

The resultant Per Capita Spent by MicroVision Segment table.

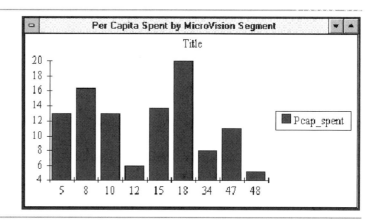

Two peaks are apparent which correspond to the MicroVision segments 8, "Movers and Shakers," and 18, "White Picket Fences." This type of chart will be useful in the future to value per capita spending in neighborhoods not covered in Tempe. Theoretically, as long as you know a neighborhood's segment, you will be able to infer its value from the analysis performed here.

Now, to generate some advertising ideas, let's explore the viewing habits of our MicroVision segments. We will create a new table from the attributes of *Mv_spent* containing only a few selected fields from the 20 records observed in Tempe.

1. Open the attribute table for *Mv_spent*. The 20 records from the earlier query should still be selected.

2. Access the Table Properties window for the attribute table by selecting Properties from the Table pull-down menu. When exporting data to a new table, only those fields checked as visible will be exported. The fields to check as visible follow:

   ❏ Blkgrp

   ❏ Prim_mv

   ❏ Second_mv

   ❏ Ave_spent

   ❏ Sum_Spent

   ❏ Count

3. Click OK to apply the above choices to the table.

4. From the File pull-down menu, select Export. From the Export Table window, select *dBASE* (the default), and click OK.

5. You will now be prompted to specify the directory and name to write the file to. Name the file *mvsel.dbf*, and click OK to export the table. The table has been created.

6. At this point, we need to add the new table to our project. Click on the Project window to make it active, and from the Project pull-down menu, select Add Table. Navigate to the directory where you wrote out the *mvsel.dbf* file, and add the table.

We are now ready to associate the above information to our cable demographics. This process is probably starting to sound familiar.

1. Copy the *Cable* theme, and change the name of the new theme to *Cable_mv*.

2. Open the attribute table for *Cable_mv*. Join the *mvsel.dbf* table to the *Cable_mv* attribute table, highlighting the *Blkgrp* field as the Join field.

3. Using the Query Builder, construct the query *Count > 0* to extract the 20 block groups of our surveyed set.

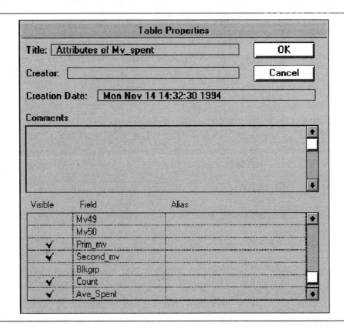

The Table Properties dialog window showing the field options for the Attributes of Mv_spent table.

Exercise 6: Working With Charts

With the MicroVision table set, we now want to create a summary table. Specifically, we want to aggregate the average values for MMI—an index which measures specific market share against a national norm—for several cable channels by MicroVision segment.

1. Highlight the *Prim_mv* field on the *Attributes of Cable_mv* table, making *Prim_mv* the summary field.
2. Click on the Summary tool to bring up the Summary Table Definition window. Select the *Cnn_mmi*, *Mtv_mmi*, *Nick_mmi*, *Usa_mmi*, *Espn_mmi* and *A&e_mmi* fields.
3. Add the above fields to the list, and then select the Average as the Summarize By option for each.
4. Save the summary table as *cablemv.dbf*.

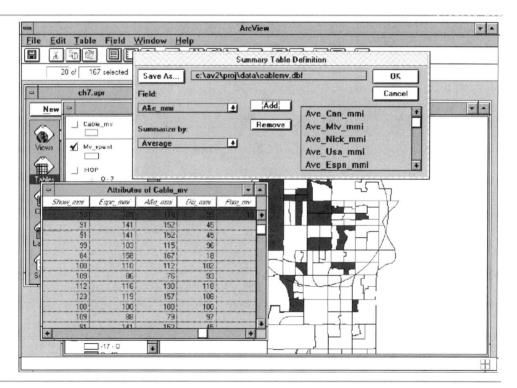

The Summary Table Definition choices for creating the cablemv.dbf table.

An output table containing nine records, one for each MicroVision segment, is created. We need the per capita amount spent by MicroVision segment, but rather than calculate it, we can simply join the table containing the per capita calculations to the *cablemv.dbf* table. To carry out the join, take the following steps:

1. Click on the Tables icon in the Project window, and open the *mvsumsp.dbf* table.
2. Highlight the *Prim_mv* field in both tables, and verify that the *cablemv.dbf* table is the destination table. (The active table is the destination table.)
3. Scroll through the *cablemv.dbf* table to examine the results of the join.

Ave_Nick_mm	Ave_Usa_mm	Ave_Espn_mm	Ave_A&e_mm	Count	Sum_Sum_Spent	Sum_Count	Pcap_spent
129.00	113.00	127.00	147.00	1	13.00	1.00	13.00
98.00	105.33	117.67	155.33	3	98.00	6.00	16.33
102.50	109.00	117.00	133.50	2	39.00	3.00	13.00
80.00	97.00	107.00	121.00	1	6.00	1.00	6.00
118.33	118.00	115.33	126.33	3	41.00	3.00	13.67
118.00	118.00	109.00	110.00	1	20.00	1.00	20.00
188.00	121.00	158.00	167.50	2	72.00	9.00	8.00
130.60	147.60	144.40	155.20	5	244.00	22.00	11.09
72.00	96.00	71.00	100.50	2	16.00	3.00	5.33

The results of joining the mvsumsp table to the cablemv table.

We are now set to chart the relationship between our MicroVision segments and their viewing habits.

1. Click on the Chart icon from the Project window, and select New to create a new chart.
2. Select the *cablemv.dbf* table to chart.
3. For Fields, select *Prim_mv, Ave_Cnn_mmi, Ave_Mtv_mmi, Ave_Nick_mmi, Ave_Usa_mmi, Ave_Espn_ mmi, Ave_A&e_mmi,* and *Pcap_spent.*
4. For Label Series Using, select *Prim_mv.* Give the chart a meaningful name, such as "Cable - MicroVision Segments." Click OK.
5. Toggle to Series Formed from Fields to obtain a more useful view if the data.

The bar chart looks promising, but this time let's switch to a line chart. Click on the Line Chart Gallery icon on the button bar, and select the second chart style from the six choices available. Click OK to create the chart. (The incremental project has been saved as *ch7c.apr*.)

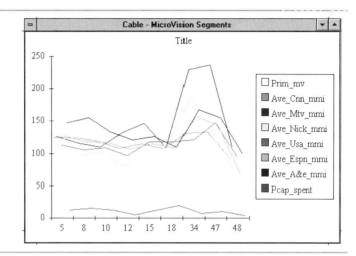

The line chart formed from the cablemv table.

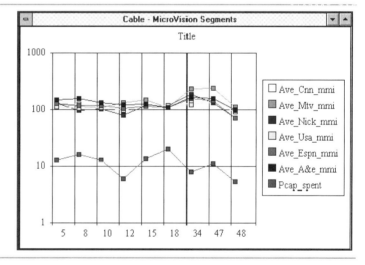

The same line chart formed from the cablemv table plotted on a logarithmic Y axis.

While we can see clear fluctuations in the cable channel MMI values, the corresponding fluctuation in *Pcap_spent* is difficult to discern. One option is to access the line chart options again, and choose the chart style on the right in which the Y axis is plotted logarithmically. This option nicely depicts the fluctuation in *Pcap_spent*, but the corresponding fluctuations in the MMI values are now muted.

166　Chapter 7: Charts

Let's take a closer look at the data. The MMI values pertain to local market share relative to the national norm for each channel. The values are percentages, with the national norm expressed as 100. What if we transform the *Pcap_spent* values to the same format?

After performing a simple calculation, we find that the average per capita amount spent is $11.20. To express each *Pcap_spent* value as a percentage of the norm, we need to divide the value by 11.2 and multiply the result by 100. To avoid two calculations for each value, you can multiply *Pcap_spent* by 8.93 (100/11.2 = 8.93).

In the following we are going to convert the *Pcap_spent* field values to the same format as the MMI values, and then update the chart.

1. Open the *cablemv.dbf* table, or click on the table to make it active if it is already open.
2. From the Table pull-down menu, select Start Editing. As before, the field names allowing editing change from italic to regular font style.

The field names on the right side of the table are still italicized because the current version of the *cablemv.dbf* table is the result of a join between this table and the *mvsumsp.dbf* table. In ArcView, all edits on tables which have been joined must be performed on the *source* table, or the *mvsumsp.dbf* table in this case.

3. Open the *mvsumsp.dbf* table, and select Start Editing from the Table pull-down menu. The field names are now shown in the normal font style.
4. From the Edit menu, select Add Field. Name the new field *Pcavg_spent*, making it a Number field, with a width of 8, and 0 decimal places. Click OK to add this field to the table.
5. Highlight the *Pcavg_spent* field, and select the Calculate tool to call up the Field Calculator dialog window.
6. Form the expression *Pcap_spent* * 8.93 by clicking on *Pcap_spent* in the Field list; double-clicking on *in the Requests list; and keying in the 8.93 value. When you have the expression formed correctly, click OK. The results of this calculation will be written to the *mvsumsp.dbf* table.
7. Close the *mvsumsp.dbf* table, and open the *cablemv.dbf* table.

8. From the Table pull-down menu, select Refresh. This action updates the joined table with any changes made in respective source tables. The new *Pcavg_spent* field will now appear in the *cablemv* table, and it contains the newly calculated values.
9. Access the Chart tool again, and select the *cablemv.dbf* table to chart.
10. Select the following fields to chart: *Prim_mv, Ave_Cnn_mmi, Ave_Mtv_mmi, Ave_Nick_mmi, Ave_Usa_mmi, Ave_Espn_mmi, Ave_A&e_mmi,* and *Pcavg_spent*.
11. Select *Prim_mv* as the field to Label Series Using.
12. Choose the line graph option. This time, both fluctuations are pronounced.

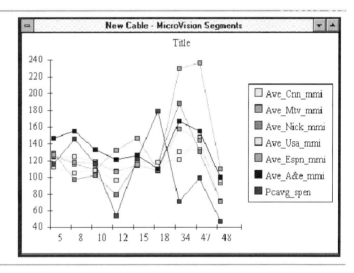

The final cablemv chart.

While no cable channel line precisely follows the amount spent line, the A&E line matches the corresponding peaks in amount spent for MicroVision segments 5 and 8 ("Prosperous Metro Mix" and "Movers and Shakers"), both of which are promising groups to target. The CNN Network, while exhibiting lower peaks, follows a similar trend. Both channels warrant consideration as possible candidates for an advertising campaign, coupled of course, with the outdoor advertising locations identified in the previous exercise.

We urge you to learn and experiment with charting. In a world where "a picture paints a thousand words," charts can be a powerful addition to your analyses and presentations.

Chapter 8

Layouts

To this point, we have worked with views, tables and charts. We have manipulated themes and linked these components together in the course of working on a project. Now we are ready to put it all together in a layout.

What Is a Layout?

A layout is a map which combines ArcView project components—views, charts and tables—along with additional elements, such as legends, scale bars, north arrows and graphics, into a single output document.

Why a Layout?

You may be asking at this point why you need to create a new document to contain what you already have displayed on your screen. Doesn't an ArcView project already do a fine job of organizing these components? Why not simply print components by selecting Print from the File menu?

To best answer these questions, we need to discuss not only what a layout is, but how it differs from other ArcView components, and how these differences translate into strengths which you can use.

A layout conceptually prepares your work for hardcopy output. It allows you to design your map to fit the format of your specific printer or plotter. You have control over the page size, the page orientation, and the output resolution.

A layout allows you to group ArcView project components and additional map elements through the use of *frames*. A frame is a container for a specific map element. A frame can contain an ArcView project component, such as a view or chart, as well as special map layout elements, such as a legend or scale bar.

A frame can be dynamically linked to corresponding ArcView project components. For example, if a view is updated in an ArcView project, the view, legend and scale bar frames will be updated as well. If a chart format is modified, the chart frame will also be updated. Frames also allow you to combine project components displayed in different windows, such as views and charts, into a single document.

A layout can also be designed to serve as a *template* for future map production. This is done by creating a layout containing empty frames, or frames which have not yet been linked to any ArcView project component. Templates can be stored for future use, at which time the links can either be established interactively, or through the use of an Avenue script, thereby facilitating automated map production.

How To Make a Layout

Begin by clicking on the Layouts icon from the Project window. Selecting New will create a default layout.

Next, you need to define the graphics page. The graphics page should correspond with the format you desire for the final output. To access the Page Setup dialog window, you select Page Setup from the Layout menu. Page parameters, including page size, page orientation, margins, and output resolution are set in the Page Setup dialog window.

Third, you can add your map elements (frames and other graphics) and link them to your ArcView project components. For ease of viewing, we recommend that you stretch the layout window as large as possible, and then fit the layout to the page by selecting Zoom to Page from the Layout menu. At this point, map elements are positioned. Frames and primitive graphics alike can be selected and moved by using the Graphics select tool. Save your work and print your map. That's all there is to it.

How To Make a Layout 171

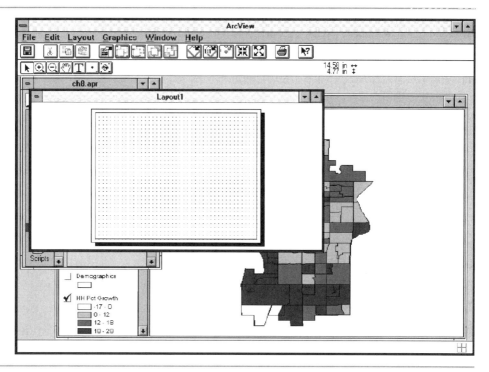

The default layout format.

The Page Setup dialog window.

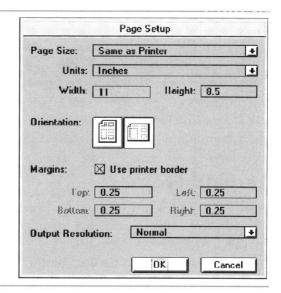

Layout Frames in More Detail

As described above, a frame is a container for a specific map element. There are seven types of frames available in a layout: view frames, chart frames, table frames, legend frames, scale bar frames, north arrow frames, and picture frames. View frames, chart frames, and table frames contain representations of the corresponding ArcView project components. Legend frames, scale bar frames, and north arrow frames contain specialized map elements, usually linked to a corresponding view frame. Picture frames contain graphics created by importing a graphics file.

To create a layout frame, click and hold the mouse button on the Frame Tool icon at the far right of the tool bar to reveal several icons for selecting a frame type. Choose the frame icon you want and then drag a box on the layout corresponding to where you want the frame located.

View Frames

A view frame contains a representation of a view from your ArcView project. The frame can either be linked to a view at the time of creation, or left empty to be linked at a later date. View frame links can either be *live* or *static*. A live link is dynamic between the view frame and the corresponding view; changes in the view will be reflected in the view frame in the layout. If the live link is not selected, the view is static, representing a snapshot of the view at the time the view frame was created.

The scale parameters of a view frame can be set as well. By default, the entire view is scaled to fit in the corresponding view frame in the layout. By checking the Preserve View Frame box, the view will be displayed at the same scale in both the view and the layout. Depending on the size of the view frame in the layout, only a portion of the view may be visible.

Legend Frames

A legend frame allows you to replicate the Table of Contents for a View on a layout. Only those themes currently drawn in the view will be included in the view legend. Legend frames will generally correspond to a view frame. They can either be linked to a view frame at the time of creation, or left empty to be linked at a later date. As with a view frame, the link can be live or static. If the link is live, the legend will be updated any time the view and associated view frame are updated. If static, the legend will stand as a snapshot of the view at the time of creation.

Scale Bar Frames

A scale bar frame creates a scale bar on your layout. Several styles are available for creating scale bars. Like legends, scale bars usually correspond to a view frame. They can either be linked to a view frame at the time of creation, or left empty to be linked at a later date. As with a view frame, the link can be either live or static. If the link is live, the size of the scale bar will change if the extent of the view is changed by zooming in or out. The size of the scale bar will also be changed if the size of the corresponding view frame is changed on the layout.

North Arrow Frames

A north arrow frame creates a north arrow on your layout. A scrolling list displays the styles of north arrows available on your system. If desired, a rotation angle can be specified when the north arrow is created.

Chart Frames

Like other frames, a chart frame contains a representation of a chart from your ArcView project. A chart frame can either be linked to a chart at the time of creation, or left empty to be linked at a later date. A chart frame is always live-linked to the corresponding chart in your project. If the chart is closed in your project, the chart frame will represent the chart with a solid rectangle containing the name of the chart. When the chart is open, the chart frame will display the current state of the chart. When the chart format is changed, the format of the chart in the chart frame will be changed as well.

Table Frames

A table frame contains a representation of a table from your ArcView project. This frame can either be linked to a table at the time of creation, or left empty to be linked at a later date. A table frame is always live-linked to the corresponding table in your project. If the table is closed in your project, the table frame will represent the table with a solid rectangle containing the name of the table.

Note that if the table is open, the table frame will display the same data visible in the table. The table will be reproduced as it appears in the project, including the shading of selected records. In addition, if the visible fields in the table exceed 80 characters in width, only the field at the far left up to a width of 80 characters will be displayed. If the displayed fields or selected records are changed in the table, the display in the table frame will be changed as well.

Picture Frames

A picture frame allows you to import graphics into a layout. The graphics could be photos, document images or work from other application software, such as spreadsheets or database forms. Imported graphics cannot be edited, although the picture frame containing the graphics can be resized.

Supported graphics formats include PostScript (including EPS), GIF, Windows Bitmap, SunRaster, TIFF, MacPaint, ERDAS, RLC, BIL, BIP, and Windows Metafile (on Windows platforms).

Layout Display Options

In addition to the many parameters described above, two display properties can be set for all frame types. The first, When Active / Always, controls when the frame is refreshed. In When Active, the frame will be refreshed only when the layout is the active project component. Always ensures that the frame will be refreshed whenever changes occur in the linked project component. The Always option allows the user to make changes in the active View window, while simultaneously viewing the changes as applied in the open Layout window.

The second display property, Presentation / Draft, controls how the frame will be displayed. When set to Presentation, the associated project component or map element will be displayed. When set to Draft, a shaded box will be displayed, representing the location of the frame on the layout, along with the name of the frame. This property can speed redraw of the screen, thereby accelerating map composition.

Map Composition

In ArcView, map composition involves creating the required frames or map elements, adding other graphics such as titles or neatlines, and arranging everything on the page until you achieve the format you desire.

Graphics

The same graphics tools available in Views (covered in Chapter 5) are available in Layouts as well. In layouts, the two most frequently used graphics tools are the Box tool for creating neatlines, and the Text tool for adding text or annotations.

Graphics manipulation tools are also available in layouts. The Group option allows selected graphics to be grouped and subsequently manipulated as a unit. Selected graphics can include frames as well as graphics primitives. This feature can greatly facilitate map composition by allowing a frame and its associated graphics, such as a title and neatline, to be repositioned as a unit.

The ability to move selected graphics to the background is another useful function. In this manner, a box can be created and shaded with a solid fill, and then placed behind a view frame to provide added emphasis.

Positioning Map Elements

Several tools exist to aid in the placement of map elements in your layout. The primary tool is the *map grid*. The map grid serves as a guide for alignment of map elements. It can serve as a visual guide, or as a *snapping grid*, to which all graphics elements will be snapped when placed.

The grid properties are controlled from the Layout Properties window, accessible from the Layout pull-down menu. The horizontal and vertical spacing of the grid can be specified, as well as whether map elements will be snapped to the grid when positioned. Also available from the Layout menu is the ability to show or hide the grid. If grid snapping is active, graphic elements will be snapped to the grid, even if the grid is not visible.

The Align tool, which you may have used for aligning graphics in a View window, is also available in the Layout window. By using this tool, selected graphics can be aligned vertically or horizontally, even if grid

176 Chapter 8: Layouts

snapping is turned off. In addition, selected graphics, including frames, can be aligned with layout margins.

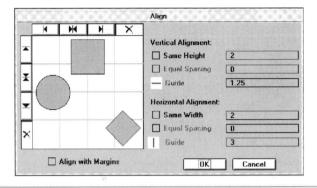

The Align dialog window.

Two methods are available to move selected graphics: positioning them with the mouse, and specifying the position relative to the layout page using the Graphic Size and Position dialog window. The Graphic Size and Position window allows you to specify a graphic element's height and width, and to position it relative to the page edges.

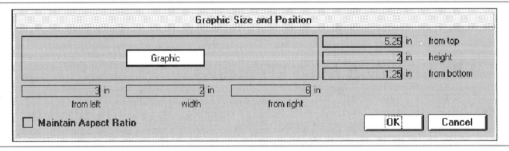

The Graphic Size and Position dialog window.

On Cartographic Design

Cartography is defined by the International Cartographic Association as the "art, science, and technology of making maps." In GIS it is safe to say that the science and technology of making maps consumes the majority of our project development time, and the creative process of design—the art of making maps—is often ignored or poorly planned.

The Five Ws of Communication

Maps are a highly specialized form of communication. A useful way to begin the cartographic design process is to answer the five basic questions in communication planning known as the five Ws:

- *Who* is your target audience?
- *What* message do you have for them?
- *Why* are you presenting this message to the target audience?
- *Where* would this message best be given to the audience?
- *When* is the best time to present this message?

Who Is Your Audience?

Marketing professionals, who spend millions of dollars trying to sell a product or service, know the strategic importance of pinpointing their audience. Even if you are not a marketing professional, you are still making the effort to communicate to a certain group of people. Who are they? What do you know about them as a group? What are their demographics? Understanding who your audience is will lend insight on how to reach them at their level.

What Message Do You Have for the Target Audience?

Perhaps you are presenting the results of an analysis project, or the conclusions to important research. Whatever your message may be, it must be clearly focused in your own mind before you can expect to effectively present it to others.

Why Are You Presenting This Message to This Audience?

What do you wish to gain from this communication? There is a purpose behind every map. It may be a reference map to navigate from A to B, or it may be to influence others and gain support for a new project. For the former, a map design which is easy to read and pleasing to behold is paramount, while the latter will benefit from a careful consideration of the psychological impact associated with the color of each map element.

Where Will You Present This Message to the Audience?

Knowing, or not knowing, the conditions under which your map will be presented will influence the design of your map. If your map is part of a larger presentation, you will want to consider the conditions under which the presentation is taking place. What type of lighting? Is the presentation room large or small? How many people will be there? What type of projection systems are available?

The media to be used influences map design. When using slides, the level of detail should be kept minimal and the short viewing time—typically less than 20 seconds—favors dramatic use of color and layout. The size and resolution of the media will affect the amount of detail which can be presented. If the map is part of a publication, it may be necessary to reduce the use of color.

When Is the Best Time to Present This Message?

We rarely have full control over the time available to create the map or for presentation. It is crucial to use the control you have over the process to your best advantage.

General Points To Consider In Design

When you are ready to design your map product, consider the following general guidelines:

- ❏ Be observant of other maps and graphic design styles. What colors work best? What text styles are easiest to read? Take note of what "works" and what you like and dislike. This will help you to develop your own style.

- ❏ Develop an organizational style, or standards that you consistently follow. If your design is strong, others will eventually come to recognize it and to associate your organization with the maps you make. This can be a very simple and effective marketing tool.

- Once you have developed a style, create ArcView templates that can be used when you have to produce maps in a hurry. Keep the templates simple, and allow for flexibility within the design to keep that creative spark alive.

Specific points to consider in designing maps and pages include the following:

- Take advantage of the power of white or negative space, that is, areas that are not part of your map. White space is a very powerful design tool.

- Build contrast between the geographic figure (the area of focus to your map purpose), and the ground (the rest of the map or the page background).

- Prioritize the geographic layers of your map based on the map purpose. What elements are most important? These elements should visually stand out more than other elements of less importance. Create a worksheet establishing this "cartographic order."

- Use one or more visual variables, such as color, shape, texture, and size, to create levels of contrast between the map elements. The point here is to create the visual illusion of the cartographic order.

- At the same time, ensure that your map consistently presents related data. Verify that features of the same type of map element are presented similarly. For example, for annotation, stick to one style per feature type. Avoid labeling similar objects with different fonts, sizes, or colors.

- Consider the entire page. Where does the map fit best considering its shape? Where will the title and legend fit best? Think in terms of the map's "flow." What does your eye focus on first, and where does it naturally move from there? Use this natural flow to guide the viewer through the logical sequence of your map "story."

- As you near completion, work through a mental checklist of the basic map elements. Have you placed a prominent title? Included an adequate legend with your name, date, and north arrow? Have you listed both map and data sources? Have you considered a "locator" map? Each of these items has the potential to contribute to and clarify your presentation.

Printing

The final step in creating a layout is sending the finished product to the printer. The Page Setup dialog box provides the initial control over what will be sent to the printer. From the Page Setup dialog box, you can set page size (standard page sizes from A through E, along with the option of entering a custom page size), page orientation (landscape or portrait), page margins, and output resolution.

> **NOTE**: Depending on your output device, graphics printed with a line weight of 1 may not correspond with the finest line possible from your printer or plotter. The remedy is to set the line weight to a fraction of 1. For graphics in a view frame, this necessitates changing the theme symbology by accessing the Legend Editor. In the box for Pen Size of Fill Outline Size, type in a new smaller value, such as 0.01. While this value will not be reflected in the line weight as drawn on the display, it will be reflected when the view or layout is printed.

Additional options are available from the Print and Print Setup dialog boxes accessed from the File menu. These options are platform-specific, and include the ability to print to a file rather than directly to an output device. Entering a file name in the To File input box will override the selected printer option. Output format can be either CGM or PostScript, and is specified by adding the appropriate file extension to the file name (e.g., .cgm or .eps).

An alternative to printing a file is to select EXPORT from the File menu. Both views and layouts can be exported. However, when exporting a view, the Table of Contents will not be exported. Supported export formats include PostScript (.eps), CGM Binary, CGM Character, and CGM Clear Text (.cgm), Adobe Illustrator (.ai), and, on the Windows platform, Windows Metafile (.wmf) and Windows Bitmap (.bmp).

Exercise 7: Working with Layouts

For those who have patiently followed along with us through the previous four exercises while we created views, manipulated themes, queried tables

Exercise 7: Working with Layouts 181

and prepared charts, the time has come to pull it all together. In this exercise we will examine layouts, the final component of an ArcView project.

In previous exercises we have used survey responses from restaurant customers to provide information on the demographics of the customer base and how the customers are distributed. In this exercise we will pull some of these elements together into a final map. Again, you will find that we have made some modifications, such as a new classification on the *Cable_mv* theme.

1. Open the *ch8.apr* project file.
2. Click on the Layouts icon in the Project window, and select New to create a new layout. A layout window containing a default layout will be opened. Resize the layout window as desired to ensure that you have room to work.
3. From the Layout window, select Page Setup, and choose the icon representing landscape page orientation.
4. From the Layout menu, select Zoom to Page. The layout will expand to fill the layout window.
5. From the Layout menu, select Properties. In the Layout Properties window, set the horizontal and vertical grid spacing to 0.2 in. The layout at this point is ready for input.

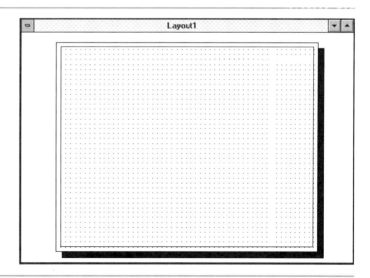

The prepared layout ready for input.

Chapter 8: Layouts

We are now ready to begin adding graphics frames, or the boxes which will contain our finished map elements, such as views and legends.

1. Select the View Frame tool from the Frame Tool pull-down, and drag a box in the lower left portion of the layout. Do not worry about the exact size and proportions because you can always resize the box later.
2. Accept all defaults in the View Frame Properties window, including the option to create this frame as an empty view. A shaded rectangle, labeled <Empty View>, will be created on the layout.
3. Add a second view frame to the right of the first in the lower center of the layout. Create this frame as an empty view frame as well.
4. From the Frame Tool pull-down, select the Chart Frame tool, and drag a box in the upper left portion of the layout. Again, accept all defaults from the Chart Frame Properties window.
5. Select the Legend Frame tool from the Frame Tool pull-down, and drag a legend frame in the lower right portion of the layout. Accept all defaults. Your layout should now look approximately like the layout in the following illustration. If necessary, use the Graphics Selection tool to resize or reposition. (The incremental project has been saved as *ch8a.apr*.)

We are not going to use the prepared layout just yet. Instead, we are going to store it as a *template* for future use.

1. From the Layout menu, select Store as Template.
2. Key in *Exercise - CH 8* for the layout name. Click OK to accept these choices.

Exercise 7: Working with Layouts 183

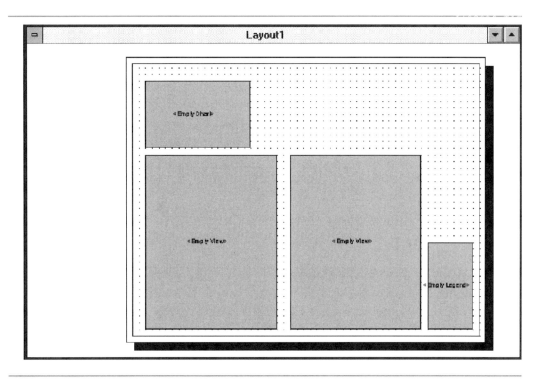

The layout with all graphics frames in place.

The Template Properties dialog window.

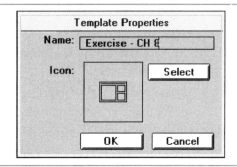

In order to use the layout template, we have some additional work to do with our views.

1. Close the layout window.

2. Create a new view window by clicking on the Views icon in the Project window. Select New. A new view window, titled *View2*,

will be opened. Move and resize this window to approximately the same size as the *View1* window.

3. Place your cursor in the *View2* window. Note that the coordinates returned in the tool bar are not the same as those returned in *View1*. This occurs because when a new view is added, it is unprojected and uses geographic coordinates by default. Before proceeding, open the View Properties window, and change the projection to Arizona State Plane coordinates (the same as in *View1*) by selecting State Plane - 1983 from the Category list and Arizona, Central from the Type list.

4. Return to the *View1* window. Click on the *HH Pct Growth* theme to make it active. From the Edit pull-down menu, select Copy Themes.

5. Return to the *View2* window, and select the Paste from the Edit menu. A copy of the *HH Pct Growth* theme is inserted into *View2*. Turn this theme on so it is displayed in the view.

We have completed our work in the views, and are ready to create the new layout.

1. Click on the Layouts icon in the Project window, and select New to create a new layout. A new layout window, titled *Layout2*, is opened. Move and resize this window as needed.

2. We are now ready to use the layout template we stored earlier. From the Layouts pull-down menu, select Use Template. The Template Manager window is opened. Click on the template titled *Exercise - CH 8*, and select OK.

3. The stored template is opened in the layout window. Note that *View1* and *View2* have been linked to the corresponding view frames. ArcView takes the initiative to link currently open view windows to corresponding view frames in a layout template. Because the legend frame could not be uniquely associated to one of the two open view windows, it was left empty. (The incremental project has been saved as *ch8b.apr*.)

Exercise 7: Working with Layouts 185

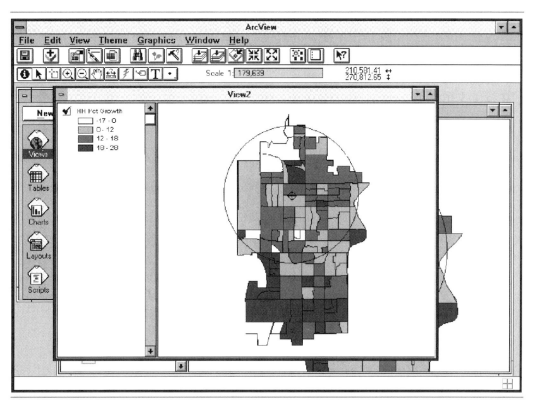

The View2 window containing the copied HH Pct Growth theme.

The Template Manager window.

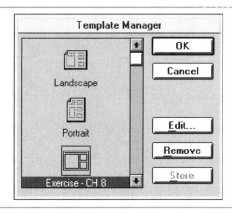

186   Chapter 8: Layouts

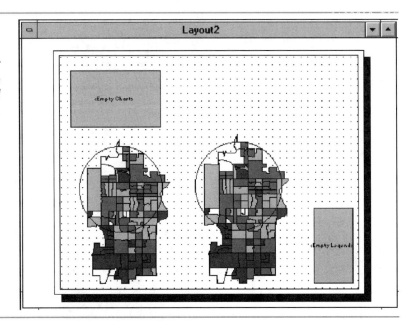

Layout2 after restoring the previously saved template.

4. With the Select tool, double-click on each frame to access the Frame Properties window. Change the display property to Draft. Note that the view frames have changed shape, reflecting the drawing area of the corresponding view windows. Reposition and resize the frame boxes according to what you feel looks best.

5. Continue by double-clicking on the legend frame, linking it to *ViewFrame2: View2*. Double-click on the chart frame, and link it to the chart titled, *Per Capita Spent by MicroVision Segment*.

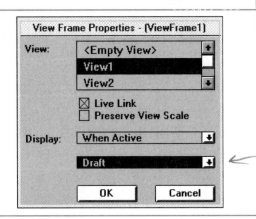

The Frame Properties window for View Frame 1.

Exercise 7: Working with Layouts 187

➽ **NOTE:** *In order for a chart or table to display on a layout, the chart or table must be open in the corresponding view.*

6. When linking is complete, the name of each project component will be displayed on the frame box.
7. Select the Box tool from the Graphics Tool pull-down, and drag a neatline around the layout.
8. Using the Text tool, add a title by clicking in the upper right area of the layout. Enter the text *Demographics of Market Survey Responses*, using two lines for text entry. Indicate that Horizontal Alignment is centered. Resize and reposition the text as needed. When complete, your layout in draft form should (or at the least *could*) resemble the layout in the following illustration.
9. Finally, double-click on each frame element, and switch the display from Draft to Presentation. If you have difficulty selecting the individual frame elements, select the neatline graphic and, from the Graphics pull-down menu, select Send to Back. (The incremental project has been saved as *ch8fin.apr.*)

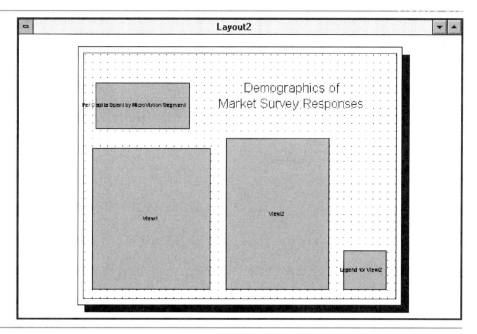

The completed layout in draft form.

188   Chapter 8: Layouts

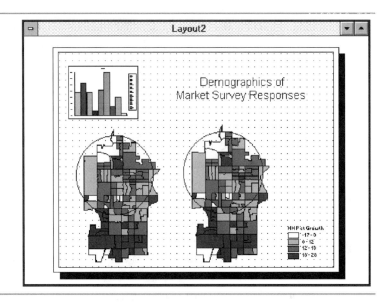

The completed layout in presentation form.

Our presentation is complete and ready to print. If you have a printer on your system, print the layout by selecting Print from the File menu.

We have now explored the four primary functional areas of ArcView: Views, Tables, Charts, and Layouts. With the use of these components alone, it is possible to design projects that accomplish a wide variety of tasks.

In the remaining chapters we explore additional functionality in ArcView, but we encourage you to visit the earlier chapters as frequently as necessary until you are comfortable with the basics. Although "basic," everything you need in order to work in the real world appears in Chapters 3 through 8.

Chapter 9

Beyond the Basics

In defining this chapter, our greatest difficulty was identifying the basics. We might agree that geocoding a table containing street addresses to create a point theme is central to many projects, but it certainly is not as basic as adding a theme from an ARC/INFO coverage. Others might argue that overlay operations such as spatial join and theme-on-theme selection which permit examination of relationships between themes constitute part of the core functionality that should be present in any desktop mapping software, and as such should be considered basic as well.

Ultimately, we elected to cover topics in this chapter which involve the *manipulation* of components already present in an ArcView project. These components are organized into the following five categories:

- Overlay Operations
- Hot Links
- Working with Shape Files
- Working with Tables
- Working with Images and Grids

Overlay Operations

Overlay operations, or spatial join and theme-on-theme selection, were discussed in previous exercises. These operations involve relating the features from one theme to the features of another. The operation can be temporary, in the case of theme-on-theme selection, or permanent, in the case of spatial join where the attributes of the source theme table are joined to the attribute table of the destination theme.

Theme-on-Theme Selection

Theme-on-theme selection involves using the selected features from one theme to define the selected set of features from a second theme, such as identifying building permits issued within a historic district. The criteria used to make this selection are determined by specifying the spatial relation type. The following six spatial relation types are available for feature selection:

- *Are Completely Within*. Features in the target theme fall entirely within features of the selecting theme.

- *Completely Contain*. Features in the selecting theme fall entirely within features of the target theme.

- *Have their Center In*. Features in the target theme have their center within features of the selecting theme.

- *Contain the Center Of*. Features in the selecting theme have their center within features of the target theme.

- *Intersect*. Features in the target theme have at least one point in common with features in the selecting theme. This includes features in the target theme that are totally contained by features of the selecting theme.

- *Are Within Distance Of*. Features in the target theme are within a specified distance of features in the selecting theme. Included are features in the target theme that are totally contained by features in the selecting theme. Specifying a selection distance effectively creates a buffer around the features in the selecting theme, although the actual buffer polygon is not visible.

Theme-on-theme selections are performed by choosing Select By Theme from the Theme menu. Pull-down lists are used to choose the selecting theme, the target theme, and the spatial relation type. If Are Within Distance Of is selected, the user inputs the selection distance in map units.

Overlay Operations

➥ *NOTE: It is possible to set the selecting theme and target theme as the same theme. If Intersect is used, this procedure can be used to locate features in a theme "adjacent" to the selected features. If Are Within Distance Of is used, the procedure locates features within a "buffer" zone of the selected features.*

When the themes and spatial relation type have been specified, the user must then choose a selection method. The following selection methods are available:

❒ ***New Set***. Creates a new set of features from all candidate features.

❒ ***Add to Set***. Adds the newly selected set of features to the currently selected set.

❒ ***Select from Set***. Selects new features from the currently selected set.

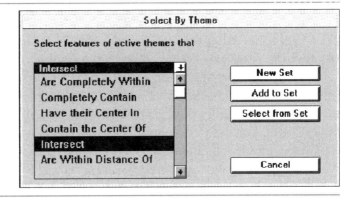

The Select By Theme dialog window.

By choosing from the different selection sets, the user controls whether to treat each selection separately, to keep a running tab, or to isolate features according to whether they meet a series of conditions. For example, target mailing lists might be generated by keeping a running list of zip codes meeting a certain criteria. Site selection might best be served by performing *select from set* operations with increasingly stringent criteria until a single zip code or two zip codes remain selected.

Spatial Join

A spatial join functions much the same way as an attribute join on two tables, with one important difference: while an attribute join is based on equivalency in attributes in the specified field between two tables, a spatial join is based on equivalency in spatial location.

The following spatial joins are supported in ArcView:

- Point to Point
- Point to Line
- Point in Polygon
- Line to Point
- Line in Line
- Line in Polygon
- Polygon in Polygon

Spatial queries can help identify customers in specified service areas, zip codes within sales territories, roads within school districts, and a variety of other data sets related to "mixed" geographies.

To perform a spatial join, open the attribute tables for the two themes, and highlight the Shape field on both tables. The table of the theme to receive the attributes resulting from the spatial join should be the active, or destination table. For example, to associate polygon attributes to point features located within those polygons, such as psychographic codes for customers typical in market research, the point theme (customers) would be the destination table. To perform the spatial join, select Join from the Table menu.

Three types of spatial relation types are used, depending on theme feature types. Point in Polygon, Line in Polygon, and Polygon in Polygon joins are performed using the Are Completely Within spatial relation type. For example, assume we wish to join two polygon themes, block groups and zip code boundaries. Block groups is the destination attribute table. The only block groups to receive zip code attributes following the spatial join are those contained entirely within a single zip code area.

A Line in Line spatial join uses Intersect as the spatial relation type, whereas Point in Point, Point in Line and Line in Point spatial joins use the Nearest spatial operator. In joins using the Nearest relation type, a

Distance field is added to the joined table. The Distance field contains the distance measured in map units between the joined features.

Hot Links

A "hot link" allows you to associate an action with a feature in a theme. The action could be to display an image or text file, open an ArcView project document such as a view or chart, or link to an external application via an Avenue script. If you have ever seen links to spreadsheets or imagery in other applications, you know that this is a powerful analysis and presentation capability.

A hot link is executed by making the theme containing the hot link active, selecting the Hot Link tool from the button bar, and then clicking on a feature in the active view. If a hot link is associated with the feature, it will execute at that time. If hot links have not been defined for a theme, the Hot Link tool will remain grayed out while the theme is active.

To define a hot link, make the theme active, then choose the Hot Link icon from the Theme Properties window to access the Hot Link Theme Properties dialog window. At this point, you specify the field from the theme attribute table which will contain the name of the file or ArcView document to access, and the action to be performed. The action is either predefined (from the choices Link to Text File, Link to Image File, Link to Document, and Link to Project), or the execution of an Avenue script.

The ability to hot link to an image file allows documents containing additional information about a feature, such as a digital photograph or scanned blueprint, to be displayed when the feature is selected. Supported image formats follow:

- GIF (Graphics Interchange Format)
- MacPaint (on Mac systems)
- Microsoft DIB (Device-Independent Bitmap)
- TIFF (Tag Image File Format)
- TIFF/LZW compressed image data
- X-Bitmap
- XWD (X Windows Dump Format)

One particular type of hot link which can be very useful is the ability to link to a specific ArcView document. This document can be a different view within the same project, making it possible to construct a "view of views." In such a project, the initial view can serve as an index to a series of views, each containing information about a specific region in greater detail.

Working with Shape Files

Shape files constitute ArcView's native format for storing spatial data. They can be created from scratch, an ARC/INFO coverage, or via translation from another desktop mapping format. Individual shape files correspond one-to-one with a particular set of geographic features. Shape files contain both graphics and associated attributes for geographic features.

The advantages to using the shape file format to store spatial data in ArcView are summarized below:

- Shape files display more rapidly on a view.
- Shape files can be edited within ArcView.
- New themes can be created in ArcView using the shape file format.
- Features can be merged or dissolved based on common attribute values.
- ARC/INFO can convert ArcView shape files back to ARC/INFO coverages.
- Shape files use an open format, allowing them to be read or written to by other software applications.

Converting to Shape Files

Any theme based on an ARC/INFO coverage can be converted to the ArcView shape file format. To carry out the conversion, make the theme active and select Convert to Shapefile from the Theme menu. If a set of active features has been selected from the theme, only the selected features will be converted to the new shape file. You will be prompted for the name and location of the shape file, and whether you wish to add the shape file to the view.

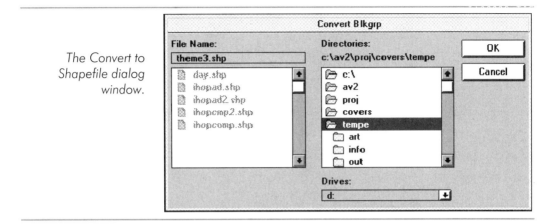

The Convert to Shapefile dialog window.

Editing Shape Files

Editing a shape file is similar to editing graphics, but with two additions: the ability to set a snapping tolerance, and the ability to merge features. To edit a shape file, make the theme active, and choose Start Editing from the Theme menu. A dashed line will appear around the theme's check box in the Table of Contents.

As with editing graphics, use the pointer tool to select features to edit. Selection handles will appear on selected features which allow you to move or resize the features. Clicking again on a selected feature will reveal the feature's vertex handles. Moving the vertex handles allows you to reshape a selected feature.

196   Chapter 9: Beyond the Basics

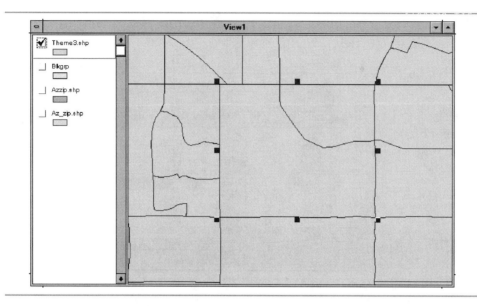

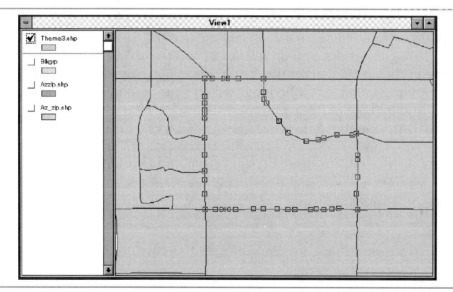

A polygon selected for editing displaying the selection handles (above), and the vertex handles (below).

ArcView imposes several limitations on shape file editing. Vertices cannot be added or deleted, and features cannot be split. You cannot add a feature type which is not present in the original theme, such as adding a polyline to a shape file containing polygons.

When you begin editing a shape file, all non-supported graphics options will be grayed out in the Graphics Tool. In addition, moving a vertex shared in common with an adjacent feature will not move the vertex of the adjacent feature. When editing polygon shapes, moving the vertex of a selected polygon will cause it to "pull away" from the adjoining polygon. To ensure that adjoining polygons are kept contiguous, turn on snapping for the theme, and set an appropriate snapping tolerance before moving the vertex of the adjoining polygon.

> ✻ *WARNING: All edits to shape files are **permanent** at the time they are made. There is no **undo**. There is no ability to quit editing without committing your edits to the theme. Exiting the project without saving will not work either. To be absolutely safe, work on a copy of your theme. Note that copying the theme in an ArcView project does **not** copy the shape file on disk: both themes still point to the same data source. A backup copy of a shape file must be made at the system level.*

While shape file editing might not be the appropriate tool for large-scale edits, it works well for small changes. Once you become acquainted with the tools, you will find shape file editing convenient for making minor edits to street networks, voting districts, sales territories, or any other type of theme you might be working with.

Theme Snapping Properties

The utility of spatial data in ArcView, from street networks to census tracts, rests in large part on feature connectivity and contiguity. Fortunately, ArcView provides the tools to help you maintain the data structure.

Turning on snapping in the theme's properties enables you to position edited vertices directly on top of existing vertices within a specified distance. If more than one vertex is found within the snapping distance, ArcView will snap to the closest vertex. Vertices edited in this manner share their coordinate location with an existing vertex, and are considered "snapped" together. Snapping can be used for editing polygons, polylines, and points.

To set the snapping properties for a theme, select the Snapping icon from the Theme Properties dialog window. A control button is displayed to turn the snapping on or off. If snapping is set to on, an input box is displayed for setting the snapping tolerance. Alternatively, the tolerance can be set interactively by clicking on the Snap tool from the tool bar, and dragging a circle on the view corresponding to the size of the desired snap tolerance. The latter method can be advantageous when you wish to visually confirm the snap distance.

> ✹ **WARNING:** *Setting the map units of an unprojected theme (a theme stored in geographic coordinates) would suggest that the same map units can be used when inputting a value for snap tolerance. This is not the case! For example, if the map units have been set to feet, setting the snap tolerance to 50 will result in a snap tolerance of 50 degrees, not 50 feet.*

Creating a New Theme

Most of the time new themes are created by selecting features from existing themes, creating a new shape file from the selected set, and editing it as necessary. Occasionally, however, you will have no choice but to start entirely from scratch. When this occurs, rest assured: whether you need to draw a few polygons to delineate sales territories, or a few lines describing alternative freeway corridors, ArcView provides the tools to accomplish this task.

To create a new theme, select the New Theme from the View pull-down menu. You will be prompted to select the feature type—point, line or polygon—and to specify the name and location for the new theme. After choosing a feature type and clicking on OK, the theme is added to the Table of Contents. By default, the theme is editable, and is ready for you to begin adding shapes.

To add attributes to the new theme, make the theme active and select the Open Theme Attribute Table icon from the button bar. The attribute table for the new theme will be opened. The table will contain the Shape field and a record corresponding to each added feature. By default, the table is editable. Additional fields can be added as needed, and attributes entered for each record. For information on editing tables, see the "Working With Tables" section later in this chapter.

Working with Shape Files

The New Theme dialog window.

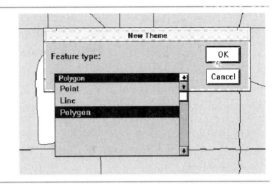

The default attribute table created for a new polygon theme.

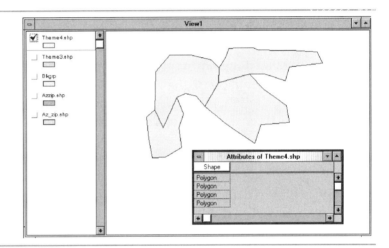

Merge Operations

Another strength of working with shape files is the ability to merge graphics into a single shape. Selecting shapes for merging can be accomplished either interactively when editing a theme, or based on a common field value in the theme's attribute table (spatial aggregation).

To merge features interactively, make the theme editable, and then use the Pointer tool or Select Features tool to select the shapes to merge. Next, select Merge Graphics from the Edit menu. The selected shapes will be merged into one shape. Note that the individual shapes do not have to be touching. In this manner, for example, individual polygons representing each of the Hawaiian Islands can be merged into one shape representing the state of Hawaii.

◆ ***NOTE:*** *When existing shapes from a shape file theme are merged into a new shape, the records in the theme attribute table for each table will be deleted and replaced with a single new record. The fields making up this new record will be blank rather than arbitrarily summarized. If necessary, the table can be subsequently edited to enter new values for the new record.*

To merge features by attribute values (spatial aggregation), open the attribute table for the active theme, and highlight the field on which to base the merge. For example, you might choose to merge a zip code theme by an attribute identifying the salesperson assigned to each zip code, thereby creating a sales territory map from the original zip code theme. In this example, you would highlight the salesperson-ID field on the zip code theme attribute table.

Next, select the Summarize tool from the button bar to access the Summary Table Definition dialog window. For the Field, select Shape, and for Summarize By, select Merge. Select Add to add this field to the summary list, and specify the name and location for the output file. Note that in the case of a merge operation, while you are being prompted for a name for the resultant output *.dbf* file, the name you provide will be used for the resultant shape file as well. When complete, select OK; the merge operation will proceed, and you will be prompted upon completion to add the new theme to the view.

The resultant output theme will contain new shapes whose boundaries correspond to unique values for the merge field from the input theme. The merge operation, most commonly associated with polygon themes, can be performed on line themes as well.

Working with Tables

In Chapter 6 we covered the basics on tables. This section is dedicated to discussion of advanced topics including editing tables, converting item types, summary statistics, and exporting tables.

Editing Tables

The initial step in editing a table is to understand the types of tables which can be edited. As you recall, ArcView can access tables in several different formats: dBase, INFO, delimited ASCII, and RDBMS tables such as Oracle or Informix via an SQL connection. However, only dBase or INFO files may be edited within ArcView. To edit other tables within ArcView, the table must first be *exported* (converted) to dBase or INFO format, and then added back into the project. (To export the table, select Export from the File menu.)

To begin editing a table, make the table active, and select Start Editing from the Table menu. If you do not have write access to the table, the Start Editing selection will be unavailable. When the table is opened for editing, the editable field names will be displayed in standard rather than italic font style.

➥ **NOTE:** *If the table is the result of a Join operation, only fields in the destination table can be edited. To edit joined fields, open the primary table and perform the edits. Use the Refresh selection from the Table menu to update the joined table after editing is complete.*

A table resulting from a join showing the editable fields in block typeface.

Shape	Blkgrp-id	Blkgrp	Count	Avg_Spent
Polygon	1450	3188.002		
Polygon	1453	3188.005	2	22.50
Polygon	1452	3188.003	2	5.50
Polygon	1479	3193.001		
Polygon	1405	3192.002		
Polygon	1487	3190.001	12	12.17
Polygon	1488	3191.001		

To edit field values, select the Edit tool from the tool bar, and then click on the cell you wish to edit. Key in the new value, followed by either <Tab> or <Enter>. The following keyboard entries are supported when editing cells:

- ❏ <Tab> moves the active cell to the right.
- ❏ <Shift>+<Tab> moves the active cell to the left.

❑ <Enter> moves the active cell down.

❑ <Shift>+<Enter> moves the active cell up.

✱ **WARNING:** *Edits to cells are written to the source table as soon as the cursor is moved to a new cell, or a new action is taken. The latter includes hitting <Esc> after keying in an error. We recommend making a backup of your data, or printing out the selected records before editing.*

On occasion you may need to add or delete a field in a table. To add a field, select Add Field from the Edit menu. You will be prompted for the field name and the field type (number, string, Boolean or date). In addition, for String fields you will be prompted for the field width, and for Number fields, the field width and number of decimal places. Keying in zero (0) for decimal places will make the field an integer. To delete a field, select Delete Field from the Edit menu.

✱ **WARNING**: *Deleting a field from a table deletes it **permanently** from the table. There is no **undo**. Removing the table from the project and adding it back in from the source will not bring it back. Again, we endorse the liberal use of backups.*

Converting Field Types

In ArcView it will become periodically desirable to change a field type in order to perform a specific operation, such as changing a number field to a string field. A change may be required by an ArcView operation demanding a certain field type, such as a number field for performing a quantile classification or generating statistics, or a string field for locating features using the Find tool. Changing a field type may also be required in order to join two tables on a common field because the field must be the same type in both tables.

The primary method to change a field type, without manipulating the table in dBase or INFO, is to add a new field to the table with the desired field type, and then converting the field as the value is calculated. To perform the conversion, begin editing the table, and highlight the target field. Select the Calculator tool to access the Field Calculator window. Form an expression from the source field, using the AsString or AsNumber request to translate the source field to the desired format.

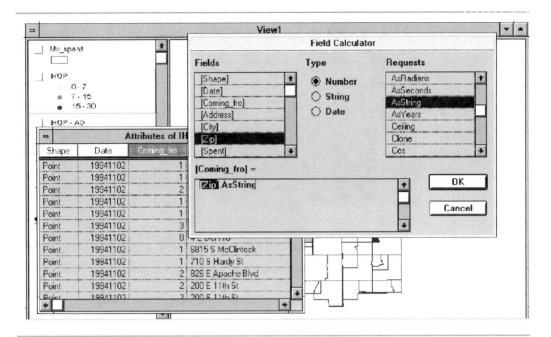

The Field Calculator dialog window showing a calculation to move the numeric field Zip to the character field Coming_fro.

Statistics

Statistics can be generated for any numeric field in an ArcView table. To generate statistics, select Statistics from the Field menu, or select the Summarize tool from the Tables button bar.

Selecting Statistics from the Field menu will create a window of statistics for the highlighted field of the selected records of the active table. If no records are selected, statistics are calculated for the entire table. Included in the output are the field's sum, count, mean, maximum, minimum, range, variance, and standard deviation.

The results of Statistics for the Pctgrowth94_99 field.

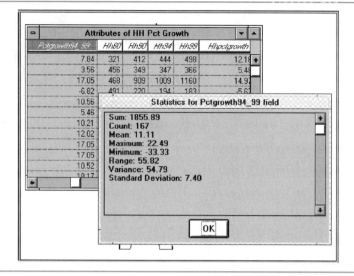

The Summarize tool is used to produce summary statistics for additional fields based on the value of a specified summary field. The results of the summarize operation are written to a new table. Summarize creates one summary record for each unique value of the summary field. The record will include the value for the summary field, a count of the records containing this value, and any summary statistics on additional fields which were requested. Available summary statistics include count, first, last, sum, average, maximum, minimum, variance, and standard deviation.

For example, a table of census demographic data could contain three fields: block number, population by block, and tract number. (A tract number identifies the census tract each block is associated with.) The Summarize tool, or summary field, can be used to obtain the total population for each census tract.

To produce summary statistics, select the Summarize icon from the button bar to access the Summary Table Definition dialog window. Two pull-down lists allow you to select the field to summarize, and the summary statistic to generate for this field. After each selection, click on the Add button to add the summary item request to the list of summary statistics to generate. When the list is complete, specify the name and directory for the output file, and click OK to create the table. The resultant view will be added to the project and opened for use.

One common use for the summary table is to join it back to the primary table using the specified summary field as the join field. In this manner the summary statistics become available for subsequent operations, such as thematic query or classification. In the above example, a quantile classification could be used to identify the most populous block groups in the study area.

> ✔ *TIP: To produce a summary table with statistics on a single field, such as those produced in the Statistics window, summarize on a field containing the same value for all records. If no such field is available, a new field can be added and left blank to serve as the specified summary field.*

Exporting Tables

Exporting a table is a convenient means of manipulating tabular data for later use in ArcView or in other applications. Export will write out all visible fields of the selected records of the active table to dBase, INFO, or delimited text format. To export a table, select Export from the File menu of the Table menu bar, and specify the export format with the name and location of the output file.

Many uses for Export may not be immediately apparent. Noteworthy Export capabilities include the following:

- ❑ Allows a custom output file to be created containing only fields and records of interest. By using Table Properties to control which fields are visible, an exported table can be confined to just a few fields and records extracted from a large table.

- ❑ Allows the results of joining tables in ArcView to be preserved for subsequent use. In this manner, additional attribute information or summary statistics can be permanently associated with records from a primary table.

- ❑ Allows data to be converted to a new format. Tables from an INFO database can be output in dBase format for subsequent use in another application. The results from an SQL query on an external database such as Oracle or Informix can be stored locally for future use.

- Allows data to be converted to a format which allows subsequent editing. Tables obtained from a delimited text file or an external SQL database can be written out locally to dBase or INFO format and subsequently edited.

If you are going to work with tabular data, it is wise to become familiar with Export.

Working with Images and Grids

ArcView provides the ability to import image and grid data into a project, as well as basic tools for manipulating this data. While a detailed description of image and grid data is beyond the scope of this book, a brief reference to selected ArcView capabilities will be useful.

Image Data

Within ArcView, image data can refer to two types of images. The first is a digital representation of a feature of interest, such as a digital photo or a scanned document. The second is a specialized digital image which represents features on the Earth (typically from an overhead view) in a format that constitutes a *digital map* of the features, such as satellite imagery and scanned aerial photography.

Aerial imagery can be a particularly powerful background to a view. ArcView supports the following digital image formats:

- BSQ, BIL and BIP
- ERDAS
- Run-length compressed files
- Sun raster files
- TIFF
- TIFF/LZW compressed image data

An image should be placed at the bottom of the Table of Contents, so that other themes will display on top of it; otherwise, the image would overlay other features. Note also that while an image represents features on the ground, it does not contain attributes about the features. Consequently, most of the operators available to feature-based themes in ArcView are not available for images. However, ArcView does provide a basic suite of tools for image manipulation, including the ability to alter the image contrast or brightness, the color map for an image, and the bands displayed in a multi-band image. Several of these features will be explored in the exercise at the end of this chapter.

GRID Data

GRID is ESRI's proprietary grid-cell, or raster-based data format. When data is converted to GRID format, the area to be mapped is divided into a series of squares with each square covering the same area. Within ARC/INFO, grids can be used to model continuous surfaces (e.g., elevation data), and complex discrete data (e.g., land use). A grid can be displayed in ArcView, and the same tools available for manipulating images work on grids. If a look-up table (LUT) has been prepared in ARC/INFO for assigning display colors to a grid (called a "colormap"), ArcView will use the LUT as well.

Controlling the Display of Images and Grids

You may encounter images and grids that cover a larger area than the other themes in the view. To prevent increased drawing time resulting from the amount of data contained in the image, the display extent of the image can be set to a value less than the full extent of the image. ArcView's display options available through the image's Theme Properties window include setting the extent to (1) the extent of all themes in the view, (2) a specific theme in the view, or (3) a user-specified display extent, using either coordinate values or the current extent of the display.

In addition, as with themes, the minimum and maximum scale can be set for the display of the image. These display properties can be set by making the image the active theme and selecting Properties from the Theme menu.

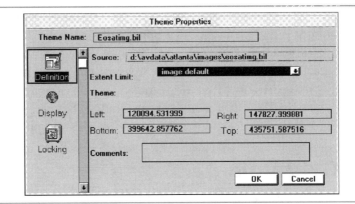

The Theme Properties window for an image theme.

Exercise 8: Shape Files and Hot Links

In this exercise we will be using sample data supplied with ArcView along with our custom data sets to demonstrate selected properties of shape files and hot links.

1. Open a new project, and then open a view window in the project. Select the Add Theme icon from the button bar, and navigate to the */avdata/namerica/usa* directory at the location where the ArcView sample data resides on your system. Add the *Usa* theme. When added to your view, this theme will appear with the default name of *Cnty*. Display the theme; you will see all U.S. counties, including counties in Alaska and Hawaii.

↦ *NOTE: If you have not previously loaded the ArcView sample data on your system, you will need to install the data at this time. An alternative is to access the ArcView sample data directly from the ArcView CD-ROM. In order to access the CD-ROM directly from the Add Theme dialog window, it is necessary to type at least the partial path in the Directory input box as opposed to selecting the drive for the CD-ROM from the Drives list. In our example, typing **f:\avdata\namerica\usa** (assuming your CD-ROM has been* *mounted as the f: drive) will result in the available data sources being listed, including the Usa cover we desire.*

2. You are viewing an ARC/INFO coverage. Because the coverage is stored in Geographic coordinates, it appears flat and distorted. With such a vast geographic area on display, as a first step we will take the opportunity to experiment with projections.

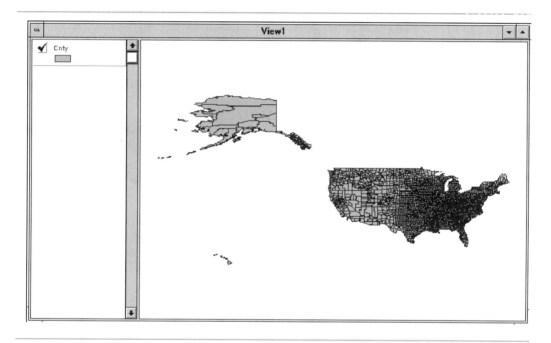

The USA County theme in geographic coordinates.

3. From the View menu, select Properties, and click on the Projection button to access the Projection Properties menu.

4. From the Categories list, select Projections of the United States. Several choices will be available in the Type list. Experiment with these choices, but remember to return to Geographic coordinates when you have finished by selecting Projections of the World from the Category list, and Geographic from the Type list.

210   Chapter 9: Beyond the Basics

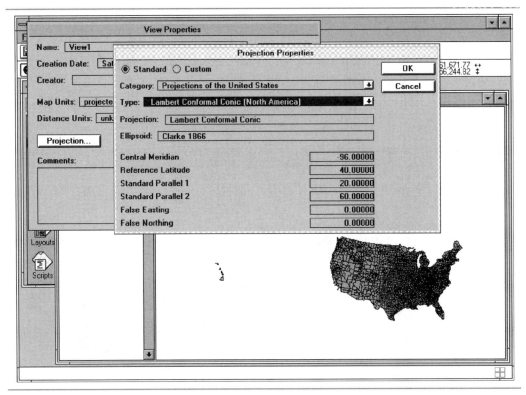

The Projection Properties dialog window as applied to View1.

At this point, you have changed the world! We will continue the exercise by zooming into Arizona to demonstrate shape files.

1. Select the Query Builder tool from the View button bar to access the Query Builder dialog box.

2. Construct the query *State_name = Arizona* and choose New Set. When complete, the selected counties for Arizona will be highlighted on the view. Use the Zoom to Selected Set icon to zoom to the selected features.

3. We can now experiment with changing the selected features to an ArcView shape file. From the Theme menu, select Convert to Shapefile. You will be prompted for the name and directory to store the completed shape file. Use the name *Az_co.shp*, and store the file in your working directory. Click on Yes when you are prompted to add the shape file as a theme in your view.

Exercise 8: Shape Files and Hot Links 211

4. The new theme will appear in the Table of Contents. Turn it on and display the results. Next, click on the *Cnty* theme in the Table of Contents, and select Delete Themes from the Edit menu. Because our selection is complete, we will no longer need the previous theme. (The incremental project has been saved as *ch9a.apr.*)

Now that we have created the *Az_co* shape file, we are ready to investigate different types of data manipulation. Let's start by examining spatial aggregation using the Merge operation.

1. Select the Add Theme tool from the View button bar, and navigate to the directory containing the data supplied with this book. Add the theme *azzip.shp*, the zip code theme for Arizona from the ArcView sample data. The theme contains an additional field, *Zip3*, the three-digit zip code class.

2. For a quick look at the distribution of the *Zip3* attribute, double-click on the *azzip.shp* theme to bring up the Legend Editor. We wish to classify the *azzip.shp* theme on the unique values for *Zip3*. Select *Zip3* from the Field list. From the Classify window, select Unique Value. Apply the classification. Your map will display the *Zip3* areas across Arizona.

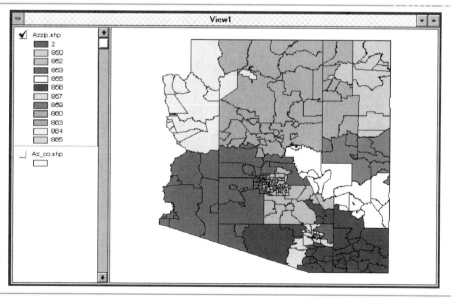

The azzip theme classified on Zip3.

212 Chapter 9: Beyond the Basics

3. We are now ready to perform Merge on the *Zip3* field. Make the *azzip* theme active, and open the theme's attribute table. Highlight the *Zip3* field on the table.
4. From the View button bar, select the Summarize tool. In the Summary Table Definition window, select Shape from the list for Field, and Merge from the list for Summarize By. Click on Add to add to the summary definition list.
5. Specify the name and directory to store the resultant output. Note that while you are prompted for the name of a file to be given a *.dbf* suffix, the Merge operation will create a new shape file. The name you supply will be applied to the new shape file as well. Use the name *az_zip3.dbf* for the output file name.

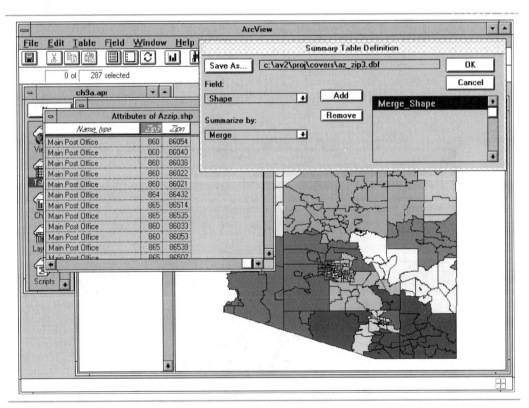

The Summary Table Definition window for the Merge operation.

Exercise 8: Shape Files and Hot Links 213

6. Click OK to begin the Merge. When complete, you will be prompted on whether to add the new theme to the view; click on OK. Display the theme to confirm your results. (The incremental project has been saved as *ch9b.apr.*)

The resultant Az_zip3 theme merged on the Zip3 field.

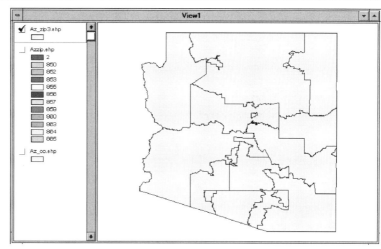

We have just aggregated all the Arizona zip code areas into larger groups, known as Arizona Zip3 areas. We are now ready to demonstrate the editing of shape files.

For the editing exercise, we are going to use the *Az_co* theme created earlier. If you run into problems, recreate the theme using the steps at the beginning of the exercise.

1. Make *Az_co* the active theme, and select Start Editing from the Theme menu. Note that the selection box for the theme in the Table of Contents is now outlined with a dashed line. Use the Zoom tool to zoom into an area in the southeast area of the state.

2. Use the Select tool and click on a polygon to edit. Note that the selection handles for the polygon are displayed.

214   Chapter 9: Beyond the Basics

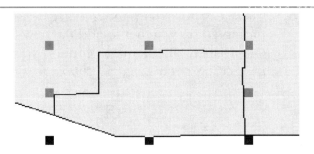

The selected polygon displaying the selection handles.

3. Click on the polygon again to reveal the vertex handles.

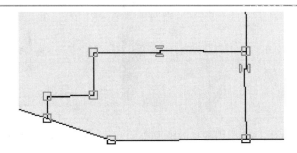

The selected polygon with the vertex handles displayed.

4. Grab a vertex with the mouse and pull it away from the adjoining polygon. Note that as the vertex is moved, the polygon is "pulled away," leaving an unshaded void.

5. Now let's move the vertex back to its original position, and ensure that it "snaps" back to its initial location. From the Theme Properties dialog window, select the Snapping icon from the bottom of the scrolling icon list. Click On to enable snapping. Leave the tolerance set to zero (0)—we will set it interactively later. Click OK to apply.

The selected polygon after moving the vertex.

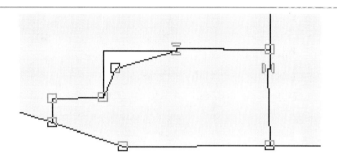

The snapping dialog box from the Theme Properties window.

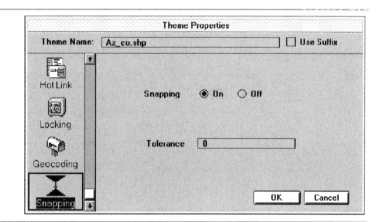

6. At this point a Snap icon has been added to the View tool bar. Click on the Snap tool to set the snapping tolerance. Place the cursor on a corner of a polygon. While holding down the mouse button, drag a circle to define the snapping tolerance. Make the circle large enough to snap the vertex when moved, but not so large as to include more than one vertex.

7. Choose the Select tool, grab the vertex, and move it back to its original position. Do not worry about moving it to the exact point—just move it to within the distance you set for the snap tolerance. When released, the vertex should snap into position. When complete, select Stop Editing from the Theme menu to end your edit session.

Chapter 9: Beyond the Basics

You have now covered the basics of editing shape files. In the next part of this exercise, we are going to create a hot link.

1. First, we need to import the project we saved at the end of the exercise for Chapter 8. Make the Project window the active window, and select Import from the Project menu. Choose *Project (*.apr)* from List Files of Type, and navigate to the directory where you have saved the chapter exercise project files. Select *ch8.apr* to import. When the import is complete, the tables and views from this project should be added to the current project.

2. Notice that if you had a view titled *View1* in the project from Chapter 8, it will be imported into the current project with the same title. This will result in two views with the title *View1* in the same project. Duplicate view titles can be confusing, but we are about to change that.

3. Click on the Views icon in the Project window. Three views will be listed: *View1*, *View1*, and *View2*. Click on the second *View1* entry, and select Open. You should see the exercise as you saved it at the end of Chapter 8. From the View Properties window, change the name of this view from *View1* to *TEMPE* (all caps). Close the view, and return to the *View1* window you have been working in for this exercise.

4. In our original *View1*, one of the fields in the *Azzip* attribute table is *Name*, which contains the city where the zip code is located. Make the *Azzip* theme active, and use the Query Builder to select the polygons for which *Name = TEMPE*.

5. Four polygons will be selected. Use the Zoom to Selected tool to zoom to the selected set. We see that four zip codes comprise the city of Tempe.

We are now ready to add the hot link. A hot link is the means in which an action can be associated with a geographic feature in a view. This action can be to display a table or image, to open another view or project, or start up an entirely new application. In this example, we are going to create a hot link which will open another view by clicking on any of the four zip codes in Tempe.

Exercise 8: Shape Files and Hot Links 217

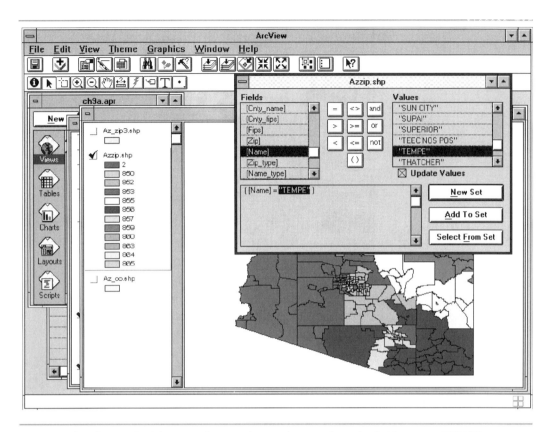

The Query Builder, ready to select polygons for which Name = TEMPE.

1. From the Theme Properties window, click on the Hot Link icon to call up the Hot Link dialog window. From the Field list, select *Name*. From the Predefined Action list, select Link to Document. This will define what happens when the hot link tool clicks on a feature for which a hot link is defined. By selecting Link to Document as the action and *Name* as the field to be used for this action, we will establish a hot link wherein clicking on a zip polygon will open a document with the corresponding city name from the *Name* field. Click OK. The Script box will show the entry *Link.Document.* Click OK again to add the hot link to the theme.

218 Chapter 9: Beyond the Basics

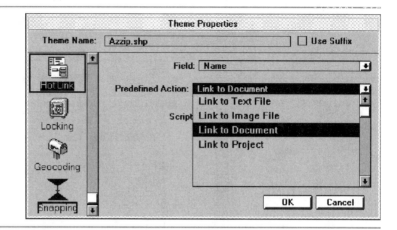

The Hot Link dialog window.

2. Click on the Hot Link tool. The cursor will change to a lightning bolt. Click on one of the four zip code polygons for Tempe. The view for TEMPE will automatically be opened.

You have just created a hot link! This particular exercise demonstrated a "view of views." By using this technique, a master coverage can serve as an index to additional themes that incorporate portions of the study area in greater detail.

Exercise 9: Working with Images

In this exercise, several tools available in ArcView for manipulating images are examined. We will use an image supplied as part of the ArcView sample data.

1. Begin by opening a new project and a new view window. Click on the Add Theme icon, and navigate to the */avdata/atlanta/images* directory at the location on your system where ArcView sample data is located.

2. Select Image Data Source from the Data Source Types list. Two images will be listed: *eosatimg.bil* and *spotimg.bil*. Select *eosatimg.bil*, and turn on the theme to display the image. You should see a predominantly black and white image of the Atlanta metro area.

Exercise 9: Working with Images   219

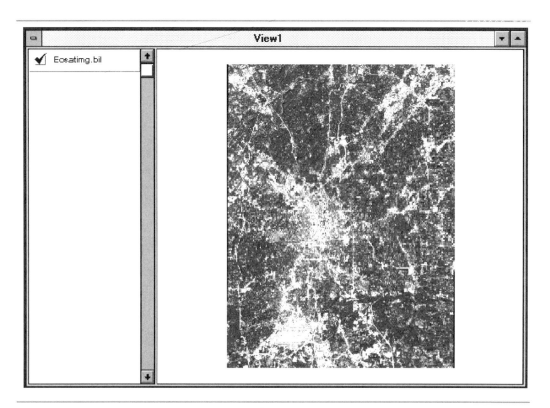

The default eosatimg image.

3. Double-click on the theme in the Table of Contents to access the Image Legend Editor. *Eosatimg* is a multi-band image. By default, the first three bands are displayed using the three primary colors for light, or red, green, and blue. A total of seven bands are available.

220   Chapter 9: Beyond the Basics

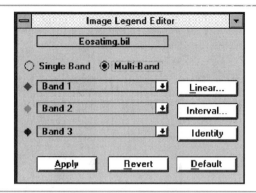

The Image Legend Editor window.

4. A useful way to examine the information contained in an image is to view each band separately. Click on Single Band to change the display from three bands to one. Note that the color used for band display changes to black. Display each band in turn by selecting the band from the list and clicking Apply. Make a mental note of what is contained on each band.

5. With band 7 displayed, click on Default. Note that the image becomes much clearer because a default linear stretch has been applied to the band. The stretch ensures that the contrast range and image brightness are set to normal values.

6. Display the other bands again, selecting Default for each band. Note the changes in the display for each band, and how additional information is revealed in what initially appeared to be a very dark image with very little detail.

7. Select band 5, click Default, and then click on the Linear button. A graph appears, which contains a curve and a straight line. The curve reveals the image's mean brightness value, and the distribution of brightness values or contrast range around this mean. The straight line with three handles superimposed on the curve serves as a tool to adjust how the image is displayed.

The Linear Lookup adjustment window.

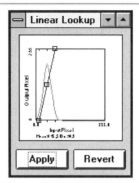

There are two rules to bear in mind when using the line to control image display:

❏ Making the slope of the line steeper or flatter increases or decreases image contrast.

❏ Moving the line to the left or the right increases or decreases image brightness.

The line is adjusted using three handles. The top and bottom handles are used to adjust the slope of the line, and the middle handle is used to move the line right or left.

8. Grab the top handle and move it slightly to the right, flattening the slope. Click on Apply. Note that the image becomes darker, but with less overall contrast. Next, grab the middle handle and move the line to the left, until the handle is located directly under the peak of the curve, and then click on Apply. Note that the contrast is still reduced, but the overall range of brightness of the image is improved.

The same linear stretch tool can be applied to a multi-band as well as a single band display.

1. Click on multi-band, select bands 1, 3 and 7, and Apply. Next, click on the Linear button. Note that three graphs are displayed, one for each band.

222   Chapter 9: Beyond the Basics

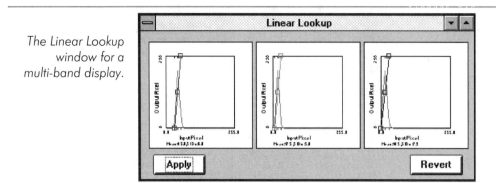

The Linear Lookup window for a multi-band display.

2. Adjust the contrast and brightness for each band, Apply, and evaluate.

In addition to adjusting the contrast and brightness for each band, the image display can also be controlled by specifying bands to be displayed in selected colors.

1. Set the red, green and blue colors to bands 7, 3 and 1, respectively.

2. Click on Default. The resultant display is often referred to as a "false color" image. In this image, band 7, which measures infra-red reflectance, is displayed in red. Band 7 in turn accentuates areas of vegetation which reflect highly in the infra-red, and thus appear red on the display.

The remaining tools in the Image Legend Editor—Interval and Identity—are used to manipulate specialized types of imagery. The Interval tool is used to control the class breaks for the display of continuous range imagery, such as digital elevation. The Identity tool is used to adjust the display of discrete class images, such as in an image classified to display vegetative types or land use. The Identity tool can also be used to manipulate the display of discrete class data stored in the GRID format.

Chapter 10

Optimizing Project Design

The previous chapters have demonstrated that ArcView provides a powerful tool for the modeling and analysis of spatial data. You probably have numerous ideas on projects you want to undertake. The only remaining issue might be one of how to begin, or how to organize project design.

We suggest that the first step is to determine the nature of your project. Is the project intended to answer a specific question, or to support a more general process? The former can be described as one-shot, or disposable applications, and the latter as reusable applications. This chapter addresses the design of reusable applications. Nevertheless, even if you are using ArcView to obtain a specific answer to a problem you do not intend to address again, the principles of project organization are relevant.

Proper database and workspace management can make it easier to work through complex assignments. These management techniques position you to more easily choose different approaches if your first efforts go wrong, and they make it easier to resume project design after interruption.

Data Organization

A successful desktop mapping project involves establishing a link between spatial data—the geographic features on your map—, and the tabular or attribute data associated with these features. How the tables are structured is highly correlated to the ease with which the data can be manipulated to answer your questions.

Spatial Data

You will typically have less control over attribute data than spatial data. If you obtain spatial and attribute data from a commercial vendor, the spatial data is determined for you based on the attribute or tabular data via a spatial or locational "hook." The hook may be a census block group or zip code. While there may be some ability to aggregate data to larger units, such as census tracts, you have little choice but to obtain the vendor's preformatted digital maps. The only situation in which you can ensure that the hook in your tabular data fits your existing spatial hooks is to generate your own tabular data via primary research.

For example, assume you obtain ten tabular data sets, and all are referenced by census tract. Meanwhile, the only spatial data set (digital map) you have represents census blocks. In this situation, you would be forced to obtain a census tract spatial data set. The underlying assumption is that replication of your tabular data set is unlikely because the data is either a snapshot in time or it would be prohibitively expensive to replicate. In contrast, the proper spatial data set can likely be obtained from a vendor or government agency, or as a last resort, digitized from scratch.

Wherever possible, verify that a one-to-one relationship exists between your mapping units and the corresponding units in your tabular data. In many instances your map may be coded with more than one locational identifier. For example, a block group theme may also be coded with additional attributes associating each block group with a census tract or zip code. While it is possible to link a table containing data gathered by census tract to this theme, a more efficient procedure would be to first prepare a new census tract theme using the Merge operation, and then link this table to the new theme. In this way, the one-to-one relationship between geographic features and attribute records is preserved. Otherwise, links to the block group theme could produce errors when attributes gathered by census tract are modeled.

Tabular Data

It is important that you organize your tabular data as well. Whenever possible, construct your tables so that no more than one join is required, that is, the join to the theme attribute table. Multiple joins increase the likelihood of an incomplete table affecting the resultant theme, particularly when files are chained and the constraining table occurs in the middle of the chain.

Data organization extends to when the data is brought into ArcView as well. A theme attribute table or joined table will often contain numerous fields pertaining to a variety of attributes. Through thematic query and classification it is possible to produce a single theme displaying a combination of attributes. However, it is usually preferable to begin with each theme representing a single attribute, and to use thematic query and overlay for exploring the relationship between themes.

Project Organization

By maintaining separate documents for storing themes, charts, tables and layouts, ArcView imposes a certain amount of organization on any project. In addition, ArcView allows you to further organize thematic data in a project by allowing multiple views.

The use of multiple views allows you to group similar data together. In this manner environmental data might be grouped into one view, demographic data into a second view, and administrative boundaries and infrastructure made common to both views. Because duplicate themes point to a single data source, there is no storage redundancy. Moreover, distributing themes in this manner reduces the size of each view's representation on the table of contents, thereby creating a potentially ideal situation where the table of contents can be viewed without scrolling.

The use of multiple views to organize a project can be aided by the use of hot links. As discussed in Chapter 9, hot links can be established between a reference theme and additional views to create a view of views, in which clicking on a feature in a reference theme can result in a view being opened to display additional information associated with a particular feature. By removing the need for the user to know how to navigate between views, greater use can be made of multiple views as a means of organizing thematic data. If the ability to navigate between multiple views is not sufficient, hot links can also be established between projects.

Project Optimization

Once the thematic data organization is determined, attention should also be given to speed and maintenance issues. In particular, when designing an ArcView application that will be used repeatedly by many people over time, these factors are just as important as data organization to the effectiveness and usefulness of the project.

Optimization for Performance

The ArcView data structure is designed to be efficient with regard to retrieval and display. Given adequate hardware and a moderately-sized project, ArcView will deliver very acceptable performance. However, projects have a way of growing in scope and complexity, and there may come a point where performance becomes an issue.

In ArcView, performance issues center around initial project startup time, and in-use performance. Optimization for in-use performance can result in improved project initialization as well.

In-use Performance

A basic method to accelerate data access and display is to *index* the data. An index can be created on one or more fields in a table. Indexing will improve operations based on values for these fields, such as simple queries, and Join or Link operations using these fields.

To create an index on a table, open the table, highlight the fields to index, and then select Create Index from the Field menu. If an index already exists for the field, the menu choice will change to Remove Index. The index will be stored in the source data directory. If the user does not have write access to the source directory, a temporary index will be created, and removed at the end of the ArcView session.

> **NOTE:** *Indexing a field will improve performance on a simple query involving the field, such as [County] = 'Lincoln'. Indexing will not improve queries involving string matching, such as [Name] = "Sm*", or queries involving comparison, such as [Age] < 18.*

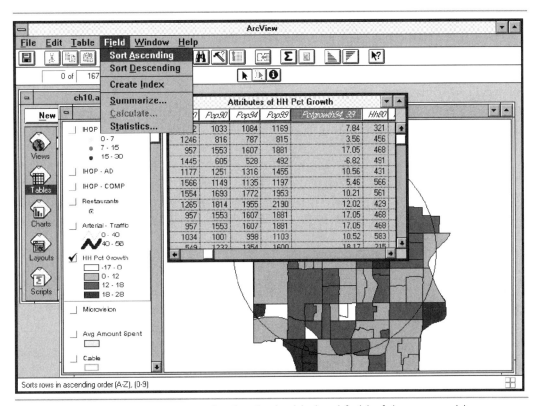

Preparing to create an index for the highlighted field of the open table.

If the Shape field is selected as the field to index, a spatial index will be created. Spatial indexes improve query and selection operations such as Identify, all drawing operations, spatial joins, and theme-on-theme selection. When a spatial join or theme-on-theme selection is performed, a spatial index is automatically created. Similar to a tabular index, a spatial index is stored in the source data directory, provided the user has write access to this directory.

Another means of improving in-use performance is to convert any complex geographic data based on ARC/INFO coverages to the ArcView *shape file* format. The shape file format is a simpler, non-topological data format optimized for use in ArcView. A theme based on an ArcView shape file will display more rapidly than a theme based on the same data from an ARC/INFO coverage. If necessary, a shape file can also be converted back to an ARC/INFO coverage for maintenance or use in other operations.

In addition, some of the methods noted in project organization can improve performance. A view-of-views can eliminate the need to display detailed themes except when zoomed into a specific area. Next, you can set a maximum display scale for detailed themes to avoid drawing when zoomed out to the full project extent. Another way to keep performance at an acceptable level is to ensure that the spatial geography is not more detailed than the thematic data it represents.

Start-up

Optimization for project initialization centers around reducing the time necessary to read the thematic data into ArcView. In order to determine the type of optimization which can be accomplished, it is first necessary to understand how an ArcView project is saved.

Because an ArcView project is *dynamic,* the project file stores *references* to the data displayed, rather than the data itself. Consequently, changes to the data made subsequent to the last ArcView session will be reflected when the project is reopened. The disadvantage here is that ArcView has to recreate all the steps taken to reach the point at which the project was last saved: all spatial and tabular data must be reloaded, all tabular joins and logical queries reconstructed, and all supporting documents, such as charts and layouts, recreated. If your project and data sources are complex, start-up time can be lengthy.

The primary means to decrease project start-up time is to reduce the number of joins and logical queries on themes and tables. Spatial and tabular data can be customized specifically for use within ArcView. This technique challenges accepted tenets of database management which stress avoiding redundant or duplicate data. However, duplicating data in order to reduce start-up time may be a worthwhile trade-off.

Logical queries on themes can be eliminated by performing the logical query once, and then using the Convert to Shapefile option to save the selected set as a new shape file. Adding the shape file to the project will make the new shape file and associated attribute table immediately available when the project is reopened. This approach may be helpful where a logical query is required to produce a subset of features based on attribute values, or where an ARC/INFO coverage contains island polygons of "no data." In ArcView these no-data polygons must be selected out to ensure that they are not subsequently shaded. If the selected set is saved as an ArcView shape file, this query will not be necessary.

The same procedure is useful for queries on tabular data. If a logical query is performed to extract a subset of records from a master data file for use in a particular project area, the start-up time required to perform this selection can be eliminated if the selected records are written to a separate file prior to opening the project.

Overhead associated with joining tables is typically not as severe as that associated with logical queries. However, when tuning for performance, reducing the number of joins to a single attribute table joined to a feature attribute table can result in improved start-up time. If necessary, all feature attributes can be added directly to the primary feature attribute table. Because of the extra overhead in coverage maintenance, however, this procedure should be considered only after all other options have been explored.

Optimization for Ease of Maintenance

Design issues involved in optimizing a project for ease of maintenance can be complex. Design criteria when the ArcView seat is part of a network, such as to a server running ARC/INFO or a network of PCs storing a distributed database, are much different than cases where the ArcView seat is located on a stand-alone PC, separate from the spatial and tabular data sources. Moreover, because optimal design for maintenance purposes may be contrary to optimization for performance, it may ultimately be necessary to balance the trade-offs.

When ArcView is used on a network linked to external spatial and tabular data sources, and performance is not an issue, the design criteria for maintenance are simple: keep as much data as possible at the source in native format. In this matter, the maintenance is retained with the originating application, and the problems associated with keeping duplicate data sets in sync are kept to a minimum. If data restructuring is required, either in post-processing of spatial data coverages or in manipulation of tabular data, this process can be performed using the originating application (including ARC/INFO), or the native DBMS software.

When the data is to be used on a stand-alone ArcView installation, data format and the ability to move data on and off the PC must be taken into account. As discussed in the insert in Chapter 3 titled, "Importing and

Exporting ARC/INFO Data," it may be difficult to establish bidirectional spatial data transfer with Windows ArcView. Consequently, you must determine whether your requirement is to update data on the PC from the source only, or to pass revised or archived data back to the host as well.

Before we provide recommendations on the proper data format to support maintenance requirements, it is necessary to review the different data formats available to ArcView. While only two spatial data formats are supported by ArcView, ARC/INFO coverages and ArcView shape files, in reality the situation is more complicated. ARC/INFO coverages come in three formats: PC ARC/INFO, UNIX ARC/INFO version 7.0, and pre-7.0 UNIX ARC/INFO. (See the insert titled, "Importing and Exporting ARC/INFO Data" in Chapter 3 for a detailed discussion of the formats.) In the discussion below we have focused on the format constraints involved in the trickiest arrangement to manage, or when moving data to and from a UNIX ARC/INFO source to a standalone Windows ArcView destination. The underlying assumption here is that you have decided you must take an active role in data maintenance.

If periodic data updates are to be provided from UNIX ARC/INFO to Windows ArcView, and the entire workspace will be updated with each delivery, then the workspace can be maintained as an ARC/INFO 7.0 workspace and copied to the PC as a unit. If updates will be performed at irregular intervals, with single coverages being updated rather than the entire workspace, then adjustments must be made to allow single coverages to be replaced on the PC without affecting the integrity of the remaining data.

If data is to be maintained in ARC/INFO 7.0 format, then each cover can be transferred to the PC in a separate workspace. Alternatively, the 7.0 cover can be converted to ARC/INFO's Export format, and restored on the PC using the Import utility. When imported, the ARC/INFO coverage is restored on the PC in the PC ARC/INFO format. However, as discussed in the "Importing and Exporting ARC/INFO Data" insert, not all ARC/INFO 7.0 coverages are supported by the Import utility. Another alternative is to convert the ARC/INFO 7.0 cover to ArcView's shape file format using ARC/INFO's ARCSHAPE command, and then copy the resultant shape file to the PC.

If updates must be passed from Windows ArcView back to UNIX ARC/INFO, there are two alternatives. If attributes only are to be edited, then the data can be maintained either as ARC/INFO coverages or as ArcView shape files. If maintained as ARC/INFO coverages, the entire workspace must be copied back to the workstation as a unit. The decision as to whether a workspace will contain a single coverage or multiple coverages will depend on whether the covers will require individual or group editing.

If both features and attributes are to be edited, the only option is to maintain the data in ArcView shape file format, the only format which allows geographic features to be edited. As indicated in the "Importing and Exporting ARC/INFO Data" insert, the Import utility creates a cover in PC ARC/INFO format. The most common way of importing this cover back to UNIX ARC/INFO is to convert the cover to Export format using PC ARC/INFO. An alternative is to convert each feature class from the PC ARC/INFO cover to an ArcView shape file, and subsequently convert the shape files back to ARC/INFO coverages using UNIX ARC/INFO.

When compared to the complexity of configuring ArcView's spatial data for optimum maintenance, managing tabular data for optimum maintenance is straightforward. ArcView supports three tabular data formats: dBase, INFO, and delimited text files. All formats can be easily transferred between the PC and UNIX workstation environments, and most database and spreadsheet software can read and write data in at least one of these formats. The watchword here is to follow standard database management procedure, that is, keep your key fields unique, and your data files discrete and modular.

Theme and Project Locking

After working on an ArcView project, it may be useful (and probably advisable) to protect the project design from being inadvertently altered. To this end, two tools are available: theme locking and project locking.

Theme locking is available for all themes in a View. Theme locking places a password on the Theme Properties window for the active theme, requiring subsequent users to issue the password to alter theme properties. This capability can be useful to protect theme properties such as a logical query applied on a theme or a hot link associated with a theme. Theme locking does not, however, lock the theme to all customization; the Legend Editor can still be invoked to change the theme classification or display.

232 Chapter 10: Optimizing Project Design

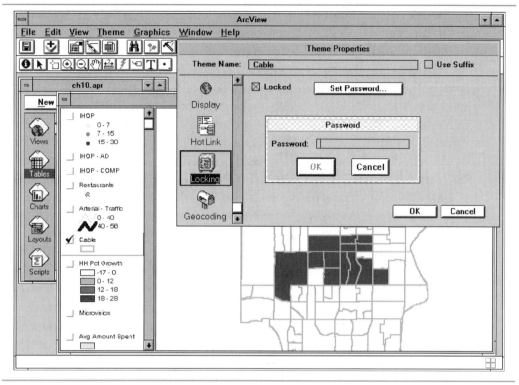

The Theme Locking option from the Theme Properties window.

Project locking can be carried out at the system level by locking the project's *.apr* file. By setting the file properties to read-only, the user can open the project and make customizations as needed, but cannot save these changes back to the original project file. If the user desires to save such changes, they must be saved to a new project file using the Save As option from the Project File menu.

Looking Ahead

As suggested above, project design and management require careful consideration even when using ArcView "straight out of the box." If you require still more control, then customizing with Avenue is your next step.

Chapter 11

ArcView Customization

Why Customize?

Through ten chapters and several detailed exercises, we have demonstrated the powers available in ArcView and how they can be used for real world problem-solving. Our focus to this point has been on using ArcView "out-of-the-box." Despite ArcView's inherent strengths, it may be necessary to customize an application to meet your needs via Avenue, ArcView's programming language. You do not have to be a programmer guru, however, to customize with Avenue. This chapter provides an overview of Avenue so that you can be aware of its strengths in the event you may someday need it.

Customizing the User Interface

An ArcView project is a collection of documents, including views, tables, charts, and layouts. Each ArcView document has a unique graphical user interface (GUI). Each GUI is composed of three control groups: the menu bar, the button bar, and the tool bar. With Avenue, the user has the ability to modify the GUI. Controls can be added, deleted or rearranged, and the properties of each control can be modified. Avenue enables you to construct a customized interface that matches user needs.

Customizing the user interface differs from the rest of Avenue functionality in that it is not necessary to write an Avenue script to achieve results.

234 Chapter 11: ArcView Customization

Instead, modification occurs through the use of the Customize dialog box which is accessed by double-clicking on any blank area on the button or tool bar, or by selecting Customize from the Project menu.

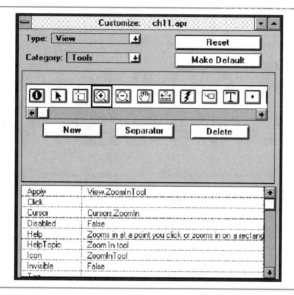

The Customize dialog box.

The Customize dialog box is divided into three functional areas. The upper area contains pull-down lists to select the document and control type to edit, as well as buttons to make the changes default or return to the standard ArcView settings. The central area, or the Control Editor, contains a representation of the control set, along with buttons to add or delete controls and separators. In the lower area, the Properties List presents a list of properties and settings for the selected control.

The choices available in the upper area's document Type list are Appl (Application), Chart, Layout, Project, Script, Table and View. The Customize dialog box allows you to modify the GUI associated with each document type. Note that the Customize dialog box allows you to modify the GUI for each document class, rather than specific instances of document types. Thus, modifications to the View GUI will apply to not only the active view, but to all view documents in the project.

The Control Editor presents the controls for the current selection—the menu bar, button bar or tool bar—as they appear on the GUI. Controls can be repositioned by clicking on the element and dragging it to a new location. The New and Delete buttons allow controls to be added or deleted, while the Separator button allows a separator to be added to the control set, i.e., a horizontal line for menu items, and a space for buttons and tools.

When a control is selected, a box is drawn around the control on the Control Editor display, and the properties of the control are displayed in the Properties List. The properties determine the behavior and appearance of the control. Properties which can be set in the Properties List appear below.

- Whether a control is visible or invisible.
- Whether a control is active or disabled.
- The icon associated with a button or tool.
- The cursor associated with a tool.
- The help string that appears in the status bar.
- The help topic associated with the control.
- The Avenue script executed by the control.

Through control of the menus, buttons and tools displayed in the user interface and the Avenue scripts associated with each control, the user is given the ability to carefully tailor the interface to fit a specific application. The ability to associate customized Avenue scripts with specific controls in substitution for the default system scripts can greatly enhance the project environment.

Saving the Customized Interface

When ArcView is started, defaults for the ArcView environment are established when the project file titled *default.apr* is read. Customization has the effect of overriding the system defaults.

The final step in customizing the ArcView interface is to make the changes permanent. Three levels of change are associated with ArcView customization:

- Changes are saved for the specific project, and no other projects are affected.

❏ Changes are applied to all projects for all users sharing the same home directory.

❏ Changes are applied system-wide.

Saving the project by using the Save button or menu selection will permanently associate the customized interface with the project. These changes will be restored when the project is opened, and take precedence over any system or home defaults.

Clicking on the Make Default button will create a new default project file (*default.apr*) in the home directory. This file will be read after the system project file, but before the application project file. The new default project file can apply to a specific user or user group, depending on whether users share the same home directory. The home default project file can be used to set general ArcView properties for a specific class of users, as well as a starting point for designing specific applications.

If the local default project file is copied to ArcView's system default project file, the changes will be applied to all ArcView users. The system default project file is also accessed by the Reset button in the Customize dialog box. Consequently, altering the system default project file makes the initial ArcView defaults permanently unavailable. The ArcView system default project file should always be backed up before making any changes.

Avenue

We are now ready to dig a little deeper and explore Avenue. The fundamental building blocks of both ArcView and Avenue are called *objects*. ArcView is built with objects, and Avenue is an object oriented programming language.

Fundamentals of Object Oriented Programming

Object oriented systems and programming languages such as Smalltalk have gained in popularity in recent years. Perhaps the best way to describe an object oriented environment is to contrast it with the more traditional procedural programming environment. In a procedural environment, such as the programming languages Fortran or C, there is a major difference between the data and the actions taken on the data. Applications constructed with procedural tools often share a strong procedural focus. Thus, in many traditional applications, such as spreadsheets and database management systems, a clear distinction is maintained between the application and the data manipulated by the application.

In an object oriented system, everything is an object. The data, application, and even the user interface are all objects in a unified environment. The distinction between the objects representing your data and the objects representing the tools used to model your data is made by defining the *properties* of each object.

The relationship between objects is defined by a formal hierarchy built on *classes*. A class is a template which defines common properties for a group of objects. Individual objects are referred to as *instances* of a class. For example, a project may contain several views, each of which is an instance of the View class.

Because classes are objects, they can in turn be associated with higher, or more general, classes. Thus, the View class is itself a member of the more general Doc (document) class. Other members of the Doc class include Chart, Layout, Table, and SEd (Script Editor). All members of the Doc class *inherit* common properties defined for that class. Individual members of the class build on inherited properties through the definition of additional properties that make them unique. Thus, while all members of the Doc class have a window and can be opened, a Table document can be sorted while a View document cannot.

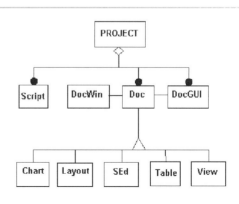

The ArcView Documents object model.

Objects interact by sending *requests* to each other. Requests are the mechanisms by which things occur in ArcView. A request can control an object, such as a request for a view to open, or return information about an object, such as a request to obtain the name of an active view. A request sent to an object in Avenue results in the return of another object. Thus, the request to obtain the name of an active view results in the return of a string object containing the name of the view.

An Avenue *statement* is comprised of objects and requests sent to the objects. The returned object can either be stored in a variable or passed on to another request. Certain requests will also accept *arguments*, such as a request to find a theme with a specific name.

Avenue Scripts

An Avenue *script*, then, is a structured series of Avenue statements organized to accomplish a given task. Several examples of Avenue scripts are discussed below.

Let's start by examining the Zoom In function. The Zoom In selection on the View menu of the View menu bar allows the user to zoom in by a determinate factor in the display of the active view. The Avenue script associated with this control follows:

```
theView = av.GetActiveDoc
theView.GetDisplay.ZoomIn(125)
```

In the first statement, the object returned by the *GetActiveDoc* request on the object *av* is assigned to the variable *theView*. (The object *av* is an Avenue reserved word which references the overall ArcView application.)

In the second statement, the object returned by the *GetDisplay* request on the active document—which returns the area of the screen which can be drawn to—is in turn passed to the *ZoomIn* request. The *ZoomIn* request takes a numeric value as its argument, which is 125 in the example above.

Thus, in a two-line script, the extent of the active view is determined and subsequently zoomed by a factor of 125 percent.

Alternatively, the user can also zoom in to the display for a view by clicking on the Zoom In tool from the View tool bar and then either dragging a rectangle describing the display extent to zoom to, or clicking on a point on which the display will be zoomed by a set factor. The following Avenue script is associated with this tool:

```
theView = av.GetActiveDoc
r = theView.ReturnUserRect
d = theView.GetDisplay
if (r <> nil) then
d.ZoomToRect(r)
else
d.ZoomIn(125)
d.PanTo(d.ReturnUserPoint)
end
```

As with the first script, the first statement returns the name of the active document and assigns it to the variable *theView*. In the second statement, the *ReturnUserRect* request—which takes as input a rectangle dragged by the user on the view—is made on the active view. The extent of the rectangle is assigned to the variable *r*.

As in the first script, the third statement uses the *GetDisplay* request to obtain the extent of the display for the active view. However, instead of being passed to the *ZoomIn* request, as in the first script, the extent of the display is instead assigned to the variable *d*.

The final statement of this script is conditional. First, the object assigned to the variable *r* is evaluated. If the value for *r* is not nil (i.e., if the user has dragged a valid rectangle), then the rectangle is passed as an argument to the *ZoomToRect* request, which acts on *d*, the extent of the current display.

If a valid rectangle was not dragged by the user, such as when the user clicks at a point on the display, then the value for *r* will be nil, causing the *else* portion of the conditional statement to be executed. In this case, the *ZoomIn* request is used, again with the argument 125, to zoom in on the current display (*d*) by a factor of 125 percent. Subsequently, the extent of the display is also centered or panned to the point clicked by the user, by way of the *PanTo* request acting on the point returned by the *ReturnUserPoint* request on the current display (*d*).

Thus, in four statements the active document is determined, and the display is zoomed either by a rectangle defined by the user or by a determinate factor at a specified point on the display. This example is a fine illustration of programming efficiency.

Avenue Syntax

By this point, you most likely have made some observations about Avenue syntax. While not a comprehensive discussion, the following will serve as a quick guide to reading an Avenue script.

As mentioned previously, programming in Avenue involves sending requests to objects. The most common format for sending a request to an object is to follow the object with the request to it by joining the two with a period or "dot" (e.g., *theView.GetDisplay*). Requests can also be chained together, such as in the following statement from the example script:

```
theView.GetDisplay.ZoomIn(125)
```

Assignment of an object to a variable is performed using the equals sign (e.g., *theView = GetActiveDoc*). If a string is assigned to a variable, the string is enclosed in double quotes (e.g., *thePres = "Clinton"*).

Although Avenue is not case sensitive, by convention variable names begin with a lower-case letter. In contrast, objects and requests begin with an upper-case letter.

Parentheses are used to supply arguments to a request. Parentheses also enclose expressions to be evaluated as part of a conditional statement, as in the expression *if (r <> nil) then* from the example script.

Comments are designated by a single quotation mark, and can be placed on a separate line or following an Avenue statement.

```
' This entire line is a comment.
theView = GetActiveDoc ' Get the active view document.
```

Creating an Avenue Script

Creating an Avenue script involves the following steps:

1. Enter the script.
2. Compile the script.
3. Debug the script.
4. Run the script.
5. Link the script to a control (optional).

Creating an Avenue Script 241

To begin creating an Avenue script, open a Script window by clicking on the Scripts icon in the Project window, and selecting New.

When a script window is the active document, the user interface is changed to the Scripts GUI. The Scripts user interface contains controls for editing, compiling and running an Avenue script. Avenue scripts can be typed directly into the Script window. Alternatively, a script can be loaded from a text file or from an existing system script.

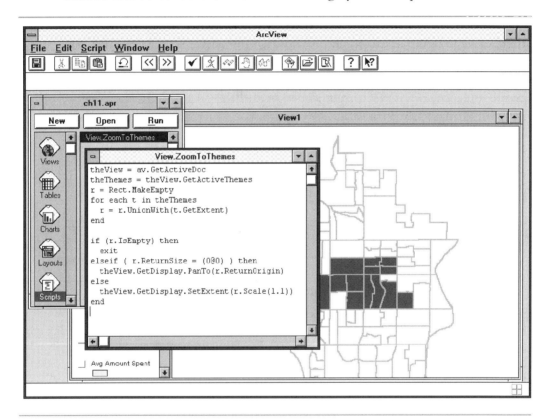

An active script window displaying the View.ZoomToThemes script.

When the script is complete, the Compile button is used to convert the script to executable format. At this point any errors in Avenue syntax are flagged. Once the script is compiled, the next step is to run the script. If the script runs but produces unexpected results, several tools are available for script debugging. These tools include the ability to step through the script one object and breakpoint at a time, and the ability to set breakpoints to interrupt script execution. The Examine Variables button is used to examine the name, class, and value of all variables at any breakpoint.

While a script can be run from the script window, typically a completed Avenue script is associated with an existing or a new ArcView control. Linking a script with an ArcView control is accomplished by calling up the Customize dialog window from the Project Properties window.

The ArcView Object Class Hierarchy

In order to program in Avenue, you must first understand how ArcView works. This may sound like a glib generalization, but the reason stems from the fact that because Avenue is an object oriented programming language, knowledge of the relationships and the structure of the objects from which ArcView is built is essential. Fortunately, the Avenue manual and the Avenue on-line help contain detailed diagrams depicting the relationships between all object classes in ArcView.

The Avenue Help System

A review of the ArcView class hierarchy can be somewhat intimidating. Apart from wallpapering your office with object diagrams, how is it possible to keep them all straight? The answer lies in Avenue's on-line help.

The on-line help system contains an entry for every object class and every request. All classes and requests are indexed for easy access using the Search utility.

Each class entry contains a description of purpose and function, several examples, and placement in the overall class structure. The overview is followed by a listing of the superclass the class inherits from, the subclasses the class is inherited by, and all available class requests and instance requests. Each listing is hyper-linked so that clicking on the name accesses the help entry for the listing.

Each request entry contains a description of what the request does, the syntax of the request, and the class of object returned by the request. Examples are provided, as well as a listing of related requests.

In practice, Avenue programming involves keeping the Avenue help window open and accessible while writing Avenue scripts. Hyper-links and the ability to search on class and request names allows for effective

Avenue programming even when one has not committed the entire object structure to memory.

The *DocGUI* listing from the Avenue on-line help system is reproduced below:

DocGUI
A DocGUI, document graphical user interface, is the collection of controls that are used to interact with a document. Each DocGUI is comprised of a menu bar, a button bar and a tool bar. Each document has a DocGUI associated with it. A project includes a list of DocGUIs defined within it. You may customize the DocGUIs delivered with ArcView or create new DocGUIs. To customize the default DocGUIs, use the Customization Panel or change the DocGUI through Avenue requests. To create a new DocGUI, clone an existing GUI or create a new one, alter its name, and add it to the project. Customize the new DocGUI through the customization panel or with Avenue scripts.
    clonedGUI = av.FindGUI("View").clone
    clonedGUI.SetName("MyView")
    av.GetProject.AddGUI(clonedGUI)
Access the components of the DocGUI through the requests
GetMenuBar, GetButtonBar and GetToolBar.
A DocGUI is associated with one or more documents; the document specifies its DocGUI with the DocGUI name. The document request SetGUI establishes the relationship. Because the DocGUI includes elements specific to the document type (for example, a scale box for views) and calls scripts which include requests specific to the document type, the DocGUI must be valid for the associated document. For example, a view must not be assigned a DocGUI designed for Tables. Use prudence to ensure the document and DocGUI are compatible when making this association through Avenue.
Inherits From
Obj
Class Requests
DocGUI
Make *(moduleName)* : *DocGUI*
Instance Requests
Accessing DocGUI Attributes
GetButtonBar : *ButtonBar*
GetCursor : *Cursor*

GetMenuBar : *MenuBar*
GetToolBar : *ToolBar*
GetType : *String*
Installing DocGUIs
Activate
IsModified : *Boolean*
SetModified *(isModified)*
Update

See also
Doc
MenuBar
ButtonBar
ToolBar
ControlSet
Application
Project

Communicating with Other Applications

Avenue can be used for a broad spectrum of tasks, from modifying ArcView controls to creating a fully customized application. Communicating with other applications warrants specific mention in this chapter, in concept if not in detail.

Communication with other applications can take two forms: sending a command string to the operating system to execute a system command or to start another application, and using *inter-application communication* (IAC) to communicate with another application in a client-server relationship.

The *System.Execute* request is a simple way to pass a request to the operating system. One obvious use for this request in the UNIX environment is the ability to execute an ARC/INFO AML (ARC Macro Language) from within ArcView. A syntax sample for this application follows:

```
System.Execute("arc \&run plotit.aml &")
```

Under Windows, a system command can be used to start another application and load a file associated with that application, such as a spreadsheet or database. For example, the following command would start Microsoft Excel and open a spreadsheet named *sales94*:

```
System.Execute("c:\excel\excel.exe
c:\work\sales94.xls")
```

IAC is a general term for a process in which two applications exchange information. In this arrangement, the process initiating the communication is referred to as the *client*, which requests data or calls functions from the responding application, referred to as the *server*. The actual protocol involved in IAC is platform dependent. In Windows, the process involves Dynamic Data Exchange (DDE); in UNIX, Remote Procedure Calls (RPC); and on the Macintosh, AppleEvent and AppleScript.

It is possible not only for ArcView to execute a command or run an AML within ARC/INFO, but also for ARC/INFO to inform ArcView when the process or AML has completed, and optionally, to return a value from the executed process to ArcView. Using the above example, ArcView could initialize an AML to buffer the ARC/INFO coverage for the active theme, and upon completion, receive the name of the new buffered ARC/INFO coverage from the AML, at which point it could be added as a new theme to the view.

Start-up and Shutdown Scripts

Every time ArcView is started, a system default file is read before any project files are opened. The *startup* file is located in the */etc* directory where ArcView is installed. In addition, a file named *shutdown* from the same directory is read when you quit ArcView.

The start-up and shutdown files are used to set the home directory for the user and display the ArcView banner. In Windows, the *startup* file will also start ArcView as a DDE Server. Upon exiting, the *shutdown* file is used to halt the DDE server.

The user also has the option to create start-up and shutdown scripts for a specific project. For example, a project start-up script could prompt the user for the path to a workspace containing ArcView data, thereby enabling a project to be made portable across platforms. Users could be prompted for their group names and passwords, after which the ArcView project specific to their group would be initialized. Such customization could enable an agency-wide application to be initialized uniformly at the system level, with the user's group used to start up a custom project tailored for that group. Similarly, a shutdown file can be used to remove any indexes created during the session, and optionally, to remove all links and joins in the event the project should be made portable across platforms.

Like modifications to the default project file, modifying the system start-up and shutdown files has the effect of globally applying these changes for all users. When modifying any system file, a backup should be made before editing so that alterations can be reversed.

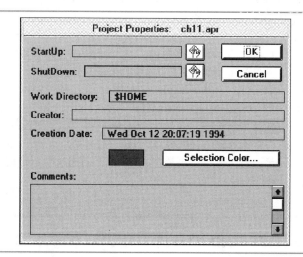

The Project Properties dialog window, with the boxes for specifying the start-up and shutdown scripts.

The Finished Application

Once an application has been completed, the final step is to make it permanent. The first step in this process may involve the creation of start-up and shutdown scripts, as described above. Additional steps could be embedding scripts, encrypting scripts, and locking the project to further customization.

Embedding a script involves taking a script stored in the Script Editor and adding it to the list of scripts stored internally with the project. Script embedding reduces the number of external documents associated with the project, thereby making the project more self-contained.

An embedded script can be viewed by any ArcView user running Avenue. However, because the embedded script will appear in the Script Manager window, the script can be loaded into the Script Editor. If protecting proprietary Avenue scripts is necessary, the script can be *encrypted*. Encryption converts the script into a format which is machine-readable only. Because this process is not reversible, the user should always store an unencrypted copy of the script in case additional changes are required.

Finally, an entire project can be locked by encrypting the project file. This action disables the ability to access the Customize dialog box or further modify the project by writing additional Avenue scripts. As with encrypting a script, encrypting the project file is not reversible.

In Summary

In keeping with the focus of this book, we have attempted to cover the basics of Avenue programming without getting too involved with details. The purpose of the above discussion was to provide information sufficient to determine if Avenue programming is required for your applications. We hope that we have provided you with a basic understanding of the ArcView data structure as well as an appreciation of how the robustness of the object oriented data model can serve to protect your investment in ArcView application development for years to come.

Chapter 12

ArcView in the Real World

After 11 chapters and much consideration, you may already have some ideas about your first few projects. Do not lose sight of the fact, however, that the applications and usage described in this text are just a few of a much broader set. Desktop mapping can impact a great many work functions and a wide variety of professions. In the interest of brainstorming on ways to use ArcView, we have compiled a few more specific descriptions of how ArcView is making contributions to agendas in the real world.

ArcView in Government

For many years government has been far and away the largest client in pushing GIS to where it is today. As the systems move to the desktop, public sector professionals continue to explore the many values of spatial analysis. Two leading examples are described below.

Arizona State Land Department

The Arizona State Land Department has been an ARC/INFO user since the software was introduced in 1982. Recently, department personnel began the development of a GIS application to aid in the management of over 9 million acres of state trust land. The GIS system links digitized state trust lease parcels to a database extracted from the Land Department's business system. Developed with the use of ArcView, this GIS system is intuitive,

Chapter 12: ArcView in the Real World

easy to use for novices, and fortified with powerful analytical and query tools for more advanced users.

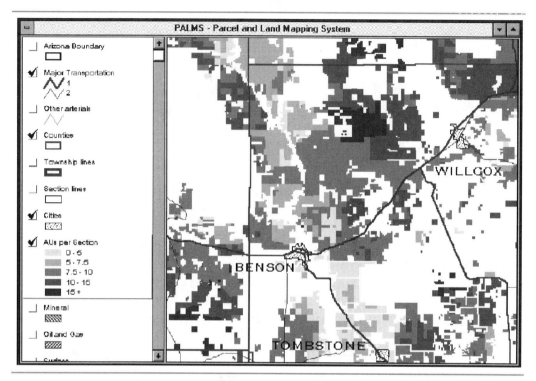

Carrying capacity for grazing on state trust land in southeastern Arizona.

The department's diverse landholdings necessitate a system that will support a wide variety of applications, from natural resource management and fire suppression to commercial property appraisal and urban land planning. Support for these disciplines is aided by the Arizona Land Resources Information System (ALRIS). ALRIS coordinates GIS activities among all state agencies and federal and local governments, and assists in making this data available to the Land Department.

Maine Department of Environmental Protection

The Geographic Information Systems Unit (MDEP-GIS) is currently developing a multi-faceted Geographic Environmental Management System (GEMS) using ArcView and ARC/INFO.

As an ArcView 2 developer site, the MDEP-GIS unit had an early start in developing the Maine Oil Spill Information System (MOSIS) for contingency planning, response, and cleanup. ArcView's object oriented architecture allows multiple modules of GEMS to be developed by sharing tools common to individual parts. An earlier version of the oil spill system developed in ARC/INFO's Arc Macro Language (AML) was conceptually redesigned to better utilize ArcView's intuitive graphic interface and object-oriented scripting language. ArcView also enables MOSIS to be used on networked PCs, thereby allowing greater portability and access to end users.

Two types of oil spill atlases are under development. The first type is the Environmental Vulnerability Index (EVI) Atlas. The EVI Atlas has two subtypes. The first set consists of Inland Fisheries and Wildlife Data, including Coastal Wildlife Concentration Areas, Shorebird Sites, Seabird Nesting Islands, and Endangered and Threatened Species, as well as the Coastal Marine Geologic Environments supplied by the Maine Geologic Survey. The second subtype is the EVI Atlas showing the Department of Marine Resources data, including Shellfish Habitats, Marine Worm Habitats, Eelgrass Beds, and Aquaculture Lease Sites.

252   Chapter 12: ArcView in the Real World

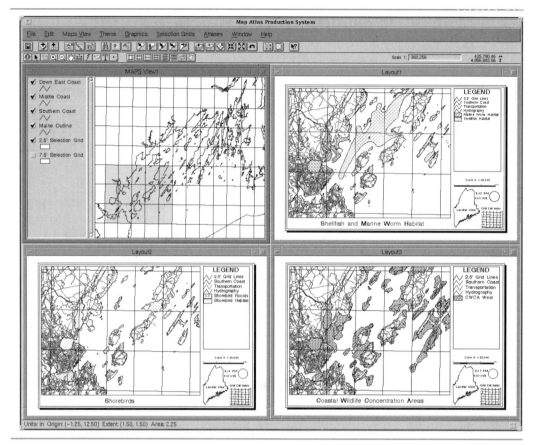

The Maine Map Atlas Production System (MAPS).

The second general type of atlas underway is the Response Logistics Atlas, containing important logistical features such as roads, water access, boat ramps, piers, fuel sources, food sources, hospitals, airports, helicopter access areas and staging areas, and a variety of additional data pertinent to an oil spill.

These systems will allow users to search or browse both geographic and attribute oil spill information and retrieve a powerful integrated set of data in a time frame inconceivable just a few years ago. The ArcView functionality and data are also general enough to support a variety of other environmental management applications.

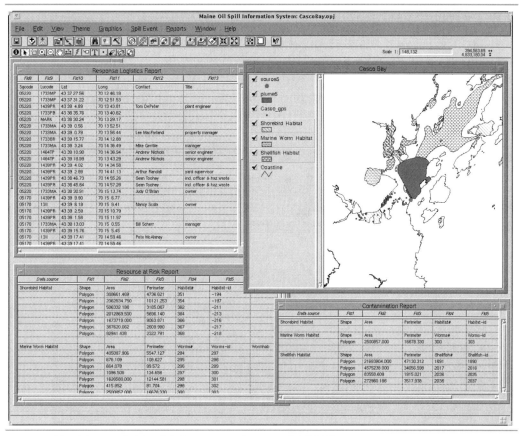

The Maine Oil Spill Information System (MOSIS).

ArcView in the Business World

While desktop mapping is still emerging and growing in stature for commercial usage, several early innovators are now directing the tools to their advantage. A general description of one of the early leaders appears below.

254 Chapter 12: ArcView in the Real World

American Isuzu

The Japanese-owned truck manufacturer has embarked on what may be the nation's most ambitious corporate program of geographic information support. American Isuzu was an early user of ArcView and is constantly seeking ways in which further deployment may enhance its operations.

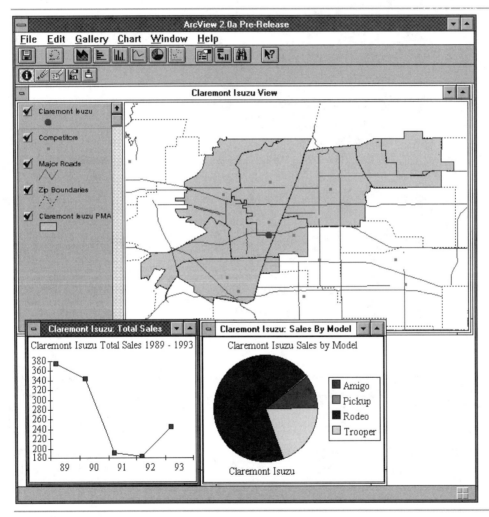

The Claremont Isuzu marketing area.

A 1994 corporation-wide needs assessment identified eight departments where desktop mapping could improve productivity or customer relations. In 1995, Isuzu implemented four particularly bold initiatives aimed at spreading ArcView throughout its sales force and advertising teams.

By year-end 1995, Isuzu expected to have a program in place for examining direct mail, pricing, warranty, and sales commission issues in all markets and dealerships within respective market areas. In the future, Isuzu intends to use GIS to facilitate and improve vehicle routing, parts marketing and training.

In an era of "one-to-one marketing," Isuzu views desktop geographic support as the competitive weapon that will bring them closer to their customers. Although Isuzu may be outsized by Ford, GM, and Toyota, Isuzu management is intent on using strategic tools like ArcView to maintain company competitiveness.

ArcView in the Academic World

Many university professors and staff regard desktop mapping as an essential research tool. In addition to the many institutions at both the middle and higher education levels that use ArcView to teach students global awareness and geographic themes, we also identified several applications that demonstrate the product's flexibility for advanced research.

University of Missouri at St. Louis

At the School of Business Administration, Dr. James F. Campbell employs ArcView to assist researchers with Washington University's Department of Anthropology to study two species of monkeys at a research station in Costa Rica. Dr. Campbell develops ArcView projects that respond to several issues, including (1) the common desire of visitors to see monkeys, and (2) the extent to which monkey observations and feeding trees are clustered in time and space.

Data was collected from a 16-month study of Ateles geoffroyi (red spider monkey) and Cebus capucinus (white-faced capuchin) at the La Selva Biological Research Station in the lowland rainforest of northeastern Costa Rica. Behavioral data were recorded at five-minute intervals on a focal animal and all trees where the animals were observed to eat were mapped with forestry flagging. These trees were then mapped to a surveyed grid. Related data was recorded in ArcView.

By his creative use of ArcView, Dr. Campbell has created projects that could help direct visitors to trails from which they might be more likely to see monkeys. For example, visitors in the summer might be directed to seek out monkeys in different locations than in the winter. ArcView is also being used to examine the density and distribution of monkey observations and monkey feeding trees relative to natural (streams) and man-made (trails) features to better understand the animals' use of space.

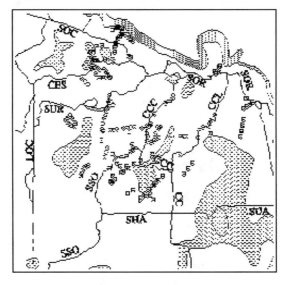

Ateles and Cebus observations in July 1991.

Ateles and Cebus observations in December 1991.

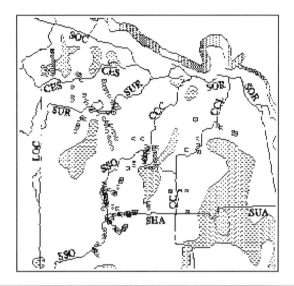

University of Wisconsin-Madison

Led by Professors Ben Niemann and Steve Ventura of the Land Resources and Environmental Studies departments, the University of Wisconsin-Madison (UWM) has assembled one of the largest academic ArcView installations in the U.S.

Among many intriguing uses of the tool has been its integration with imagery and global positioning systems (GPS). The UWM team has taken ArcView to south Florida for evaluating the recovery of natural areas from hurricane damage. Using large-scale aerial imagery as a backdrop, they have marked trails through hurricane damaged areas by kinematic GPS. Information on the research plots, aerial photography, and facilities is quickly recalled through the ArcView interface.

The UWM group has perhaps been most active with environmental concerns. Working in conjunction with the Nature Conservancy, for example, they have generated geographic data and ArcView software to drive a songbird habitat model for the "deepwoods" songbirds of the Baraboo Hills in southern Wisconsin. Using their system, researchers can evaluate areas of good, moderate or poor songbird habitat and the corresponding songbird viability. Without reforestation, the model predicts songbird numbers which are stable or in decline over time. "What-if" scenarios can then be run to determine whether reforestation will have a positive impact on songbird populations.

The team has also been a notable supporter of Marquette, Michigan's efforts at better managing stormwater quantity and quality. Working with the University's Land Information and Computer Graphics facility, the city has developed GIS-based methods to assess fees for individual landowners that are proportional to their stormwater contribution. ArcView is used to recall information on parcel size, impervious area, land use, zoning and other related data for this stormwater utility.

Larry and Penny Thompson Park biological monitoring plots.

Snapper Creek Hammock bike trails.

> **NOTE:** *ESRI has been actively promoting the use of its products in academic curriculums. If you have an interest in this area, we encourage you to contact the company's office in Redlands, California.*

Elsewhere

The world of desktop mapping and ArcView seems to be expanding daily. Scan the ESRI User Conference agenda and you will find someone reporting on nearly every aspect of life. From business to agriculture, municipal services to defense, economics to environmental studies, ArcView has something to offer.

We encourage you to attend the annual ESRI User Conference to explore more ideas with ARC/INFO and ArcView, as well as other national mapping conferences that demonstrate the power of desktop mapping. Local groups can also be helpful in expanding your thinking on the range and methods of tasks that can be performed.

Let the rest of the mapping community know about your work. Inquire about presenting at the conferences by contacting the editors of pertinent professional journals or by contacting your local ESRI representative.

Glossary

Address Events

An address event is a feature located by a unique address. ArcView-supported addresses include street addresses (the most common), as well as polygon and point addresses, such as zip codes or land parcel numbers, respectively. (See also: *Event Table, Polygon.*)

Address Matching

Address matching involves assigning an absolute location through x,y coordinates to each address in an address event table. This process occurs through interpolating specific address locations against a geocoded street theme coded with address ranges. (See also: *Event Table.*)

Alias Table

ArcView allows an alias table to be used for address geocoding. In an alias table, you can use place name aliases in which a place name, such as Madison County Hospital, is associated with a street address. The alias table is associated with an address event table in address geocoding, or with a matchable theme to facilitate locating addresses in a view. (See also: *Event Table.*)

Attribute Data / Table

Attribute data, also known as tabular data, are linked to themes. ArcView shape files (and ARC/INFO coverages) contain spatial data and attribute tables. The attribute table(s) linked to a particular theme may contain geographic information (e.g., addresses or zip codes). Attribute tables can also contain information associated with features in a theme, such as soil properties or land use descriptions.

   A one-to-one relationship exists between features in a shape file (or coverage) and records in the theme attribute table. At a minimum, the theme attribute table contains a shape field. In addition, theme attribute tables derived from coverages contain an additional field containing a unique numeric identifier for each feature. Fields may be added to the theme attribute table to identify additional characteristics of features. An example would be a block group number associated with each polygon of a census geography theme. Fields can also be used to join additional attribute data to the theme attribute table. (See also: *Field, Join, Record, Table.*)

Chart

A chart is a graphic presentation of attribute or tabular data. In ArcView, a chart references data from a table in a project in one of six formats: area charts, bar charts, column charts, line charts, pie charts, and xy scatter charts. An ArcView chart is dynamic, in that the chart represents the current status of the data in the table. Changes to either data values or the selected records in the table are immediately reflected in the corresponding chart. (See also: *Dynamic.*)

Continuous Events

A continuous event is a type of route event. Continuous events are features located in a continuous fashion along a route. Assume the route system is a natural gas distribution network. The continuous event could be pipeline age organized into categories such as very old (installed before 1965), old (1965 to 1975), moderate (1976 to 1985), new (1986 to the present). The gas distribution network would be coded according to where pipeline age changes. (See also: *Route Events.*)

Coordinate System

A coordinate system is a map reference system in which precise geographic position can be referenced for a local area by means of a rectangular grid. The use of a rectangular grid allows features to be located using x,y coordinates. This system facilitates the integration of survey data into a larger national grid.

Each coordinate system is derived from a specific map projection. The Universal Transverse Mercator System is derived from the Transverse Mercator projection. The Universal Transverse Mercator System divides the world into 60 north-south zones, and 20 east-west zones, for a total of 1200 unique grid zones.

The State Plane Coordinate System is used in the United States. In this system, each state is divided into one or more zones extending either north-south or east-west. Zones extending north-south are based on the Transverse Mercator projection, while zones extending east-west are based on the Lambert Conformal Conic projection. (See also: *Geographic Coordinates, Map Projections.*)

Coverage

In the context of this book, coverage refers to an ARC/INFO coverage. An ARC/INFO coverage is a database that stores geographic and tabular data in a set of files. These data files are organized within a common directory. An ARC/INFO coverage can serve as a spatial data source for a theme in ArcView.

Data Group

A data group is the aggregating unit of a chart. It consists of a set of related elements that describe the same variable. If the data series is formed from records, the data group is aggregated by fields. If the series is formed from fields, the data group is aggregated by record. (See also: *Chart, Field, Record, Table*; insert titled, "Clarification of Data Markers, Series, and Groups.")

Data Marker

A data marker is an element on a chart--a column, bar, area, pie slice, or point symbol—which represents the value of a particular field for a specific record in a table. Data markers are analogous to cells in a spreadsheet. In other words, the data marker in a chart represents the intersection of a field and record in a table, known as a field value. (See also: *Chart, Field, Record, Table*; insert titled, "Clarification of Data Markers, Series, and Groups.")

Data Series

A data series is the set of values that are compared in a chart. The individual elements of the data series, which may be formed from the records or fields in a table, are displayed in the chart legend. (See also: *Chart, Field, Record, Table*; insert titled, "Clarification of Data Markers, Series, and Groups.")

Datum

A datum is a reference system used to describe the surface of the Earth. Coordinate systems used in surveying to locate points on the Earth, such as the State Plane Coordinate System, or the Universal Transverse Mercator Grid, are linked to a specific datum. For North America, there are two: the North American datum of 1927 or the North American datum of 1983. (See also: *Geodetic Control*.)

DOS 8.3 Convention

Under DOS, file names are limited to a maximum of eight characters, followed by a maximum three-character extension. The field name and extension are separated by a period. Examples of the convention are highways.dbf, and marker.ai. Directory names are also restricted to the same naming convention.

Dynamic

In ArcView, dynamic refers to a document (e.g., table, chart, or layout) which reflects the current status of the source data it is based on. As the source data changes, the document dynamically associated with the source data reflects the changes.

Event Table

An event table contains geographic locations ("events"); the table is not in a spatial data format. The geographic locations may be absolute (e.g., latitude and longitude coordinates), or relative (e.g., street addresses). Event tables contain of one of three general types of events: x,y, route or address. (See also: *Address Events, Route Events, X,Y Events*.)

Event Theme

An event theme is a theme based on an event table. Event themes are based on x,y events, address events, or route events. When the geographic locations are absolute, as in the case of with x,y events, points are created in a theme directly from the x,y coordinate values. When the geographic locations are relative, feature locations are translated from relative to absolute locations, and the resultant features stored in the ArcView shape file format. A theme referencing this shape file is subsequently added to the active view. (See also: *Event Table*.)

Export

This term is used in both a general and very specific sense in this book. The general definition is simply the process of moving a file prepared in ArcView to another application.

Next, Export is a utility provided with PC or UNIX ARC/INFO which creates an ARC/INFO interchange format file. The ARC/INFO interchange format file is given the suffix Enn (where nn is a number from 00 to 99) for each successive volume created.

Field

A field is a unique descriptor or characteristic of a record (instance) in a database. Fields are also called columns or items. In a customer database, examples of fields could be name, address, city, and zip code. ArcView supports four field types: Number, String, Boolean, and Date. (See also: *Record, Table*.)

Geocoding

Geocoding is the process of assigning an absolute location (x,y coordinates) to a geographic feature referenced by a relative location, such as a street address or zip code.

Geodetic Control

Geodetic control consists of a correlated network of points for which accurate elevation and position locations have been determined. Local surveys are subsequently adjusted to such a correlated network of points. Control points are tied into existing horizontal and vertical control networks, and then adjusted to the appropriate datum. The appropriate datum for North America would be the 1927 datum or the 1983 datum. (See also: *Datum, Map Projections.*)

Geographic Coordinates (Latitude/Longitude)

Geographic coordinates refer to the geographic reference system of latitude and longitude in which the Earth is treated as a sphere, and divided into 360 equal parts (degrees). This division is performed along two axes, one running east-east along the equator, and the other running north-south along the Greenwich Prime Meridian. Using this coordinate system, any location on the Earth can be identified with a unique x,y coordinate pair. Geographic coordinates are commonly measured in degrees, minutes, and seconds, and can also be formatted as decimal degrees. The format used by ArcView is decimal degrees. (See also: *Coordinate System, Map Projection, Map Units.*)

GIS (Geographic Information System)

A GIS is a geographic database manager. In other words, a GIS treats all geographic (spatial) features as records in a database, and not simply as graphics. Nearly all concepts in traditional relational databases apply to GIS, but with the added dimension of geography.

A GIS builds a bridge between geography and descriptive information through a georelational model. This model means that there is a one-to-one relationship between a spatial data set and an attribute table. Some database fields are predefined for you in attribute tables, but you can add any fields you desire. The georelational model also permits you to connect to other tabular databases, whether internal or external to the GIS software.

Hot Link

A hot link is a user-defined action that occurs upon selecting the Hot Link tool from the button bar, and then clicking on a feature in a theme. The user defines the action executed at this point through specifying the value of a field from the theme attribute table, and the action to be performed. Examples of hot link actions are displaying an image or opening another view.

Image Data

Image data are a form of raster data in which features have been converted to a series of cell values by an optical or electronic device. Image data typically refers to satellite imagery or scanned aerial photographs in which each grid cell contains a brightness value representative of the portion of the light spectrum being measured. Imagery may be single band (gray-scale), or multiple band (multispectral). ArcView supports the display of both single band and multiple band imagery. (See also: *Raster Data*.)

Import

This term is used in both a general and very specific sense in this book. The general definition of importing is simply the process of loading or retrieving data into ArcView. Examples are importing dBase or delimited text files into ArcView. Both of these file types were created outside of ArcView, and then incorporated (imported) into the program.

Import is a stand-alone utility supplied with ArcView that allows an ARC/INFO interchange file to be converted to an ARC/INFO coverage or database data file. The Import utility allows ARC/INFO interchange format data to be converted to a format which can be added to an ArcView project. Import creates a PC ARC/INFO format coverage under the Windows platform, and a workstation ARC/INFO coverage under UNIX. The Import utility also allows project files within ArcView 1 (*.av* files) to be imported into an ArcView 2 project.

Join

Join is a process by which two or more tables are merged in a virtual table through a common field. The resultant table appears as a single table in ArcView, despite being formed from two or more separate source data files.

Join is used when there is a one-to-one or one-to-many relationship between records in the source table and records in the destination table.

In the case of a many-to-one relationship between records in the source table and records in the destination table, Link should be used instead of Join. (See also: *Field, Link, Table.*)

Layout

A layout is a map composition document used to prepare output from ArcView. A layout allows the user to define the graphics page and place ArcView documents (views, charts, and tables), imported graphics, and graphics primitives on the page. A layout is dynamic, in that graphics linked to ArcView documents can immediately reflect changes in respective documents. Dynamic layouts are live-linked. (See also: *Live Link.*)

Line Features

Line features are used to represent linear entities, such as highways or streams. Other important uses for lines are to delineate perimeters of polygons and to lay down the positions of route systems and regions. Lines are located are defined through the assignment of a unique series of x,y coordinate pairs.

Lines (also known as arcs) are connected strings of line segments. Each line segment is delineated by a vertex. The vertices at the endpoints of lines are called nodes. (See also: *Vertex.*)

Linear Events

A linear event is a type of route event. Linear events are features located along a specific segment of a route. Assume the route system is a road network. A linear event could be a pothole repair segment occurring from milepost 10.5 to 10.8 on County Road 41. (See also: *Route Events.*)

Link

Link is used in the case of a many-to-one relationship between records in the source table and records in the destination table. Link defines the relationship between the two tables, but does not bring the tables into a single virtual table. Link displays all candidate records in the source table matching each unique value in the destination table. For example, a table identifying apartment complexes by parcel number could be linked to a second (source) table listing all the tenants at each apartment complex. (See also: *Join.*)

Live Link

A live link is a dynamic link between a layout frame and the corresponding ArcView document or element. Chart and table frames are always live-linked to respective chart and table documents. View, legend, and scale bar frames can be either live-linked or static. (See also: *Dynamic, Layout, Static Link.*)

Look-up Table

A look-up table is a secondary table containing additional information about features identified in a specific field of the primary table. For example, the primary table may contain a list of soil mapping units, including a field containing a symbol for each mapping unit. A look-up table could then be prepared that identifies the soil drainage class for each soil mapping unit.

A one-to-one or many-to-one relationship exists between records in the primary table and records in the secondary table. In ArcView, Join can be used to create a virtual table in which the records of the look-up table are associated with records in the primary table based on values for a common field.

Map Projection

A map projection is a system by which the curved surface of the Earth is represented on the flat surface of a map. The problem inherent in all map projections is preserving the properties of area, shape, elevation, distance and direction present on the Earth's surface through transformation to a map surface. Because it is impossible to preserve all properties simultaneously, it is necessary to select a map projection optimized to preserve the property most desired.

Map Units

Map units are the units in which spatial data coordinates are stored. Map units are used in ArcView to set the scale of the view. By default, the map units for the are set to Unknown. The map units for a view are set from the View Properties dialog window which is accessed from the View menu.

Map units are differentiated in ArcView from distance units. Distance units are used to display measurements and dimensions within a view.

Matchable Theme

A theme becomes matchable after geocoding indexes have been built on it that conform to an ArcView-supported address style. A matchable theme is used to create an Address Event Theme by address matching from a table containing address information. (See also: *Event Theme*.)

Point Events

A point event is a type of route event. Point events are located at a specific point along a route. Assume the route is a highway system coded by route (e.g., Route 41 linking Tijeras Canyon to the town of Milbank) and mile post. Point events could be accident locations identified by a specific route mile post on the route system. (See also: *Route Events*.)

Point Features

Point features represent entities found at discrete locations, such as well sites, transformer sites, or customer locations. Each point feature is located using a single x,y coordinate pair.

Polygon Features

Polygon features are used to represent entities of a real extent, such as land parcels, geologic zones or islands. Polygon features are defined by a series of x,y coordinate pairs identifying the polygon's perimeters.

Imagine four discrete points (x,y coordinate pairs), and four lines (arcs) connecting the points in a closed system. The points and lines define the polygon's perimeters. The entity (area) thus defined is a polygon feature.

If polygon features are derived from an ARC/INFO coverage, a special type of point feature known as a label point may be associated with each polygon. The label point contains the same attributes as the polygon, and enables the polygons to alternatively be modeled using point features.

Project

A project is the overall structure used in ArcView to organize component documents such as views, tables, charts, layouts, and scripts. A project maintains the current state of all component ArcView documents, as well as the configuration of the graphical user interface. This information is stored in an ASCII file with an .apr extension.

Raster Data

Raster data is also referred to as grid-cell data. Raster data sets store spatial data as cells within a two-dimensional matrix. This matrix consists of a gridded area of uniformly spaced rows and columns in which each grid cell contains a value representative of the feature being depicted. Grid cell values may be continuous, as in elevation data, or discrete, as in land use data where each grid cell value is associated with a specific land use. The resolution of raster data is dependent on the size of the grid cell.

Record

A record is a specific instance or member of a database. Records are also called rows. In a customer database, records could be individual customers. (See also: *Table*.)

Route Events

Route events are relative locations of features along a route system, such as a road network or power lines. Locations are relative in that they are referenced as distances from a known starting point, such as 1.2 miles from the Salt Creek substation. Route events come in three forms: point, linear and continuous. (See also: *Continuous Events, Event Table, Linear Events, Point Events*.)

Script

An ArcView script is a macro written in Avenue (ArcView's programming language) to customize the ArcView environment. Avenue scripts are stored and executed within a project.

Shape File

A shape file is the native ArcView spatial data format. In contrast to an ARC/INFO coverage, a shape file is a simpler, non-topological format which offers the advantages of faster display and the ability to be created or edited within ArcView. An ArcView shape file also serves as an effective interchange format for moving data in and out of ARC/INFO or other supporting software.

Static Link

A static link is a non-dynamic link between a layout frame and the corresponding ArcView document or element. If a view, legend or scale bar frame is static, the frame represents a snapshot of the view at the time the frame was created. (See also: *Layout, Live Link*.)

Street Net

A street net or network is a spatial data set representing streets as a connected series of line segments. Each line segment is typically coded with the address range of the street represented, thereby allowing the theme resultant from the street net to be used for geocoding.

Table

A table is the basic unit of storage in a database management system. It is a two-dimensional matrix of attribute values. ArcView, ARC/INFO and SQL (Structured Query Language) share this term.

Fields are the vertical components in a table. In SQL, fields are also called columns. Another common term used for fields is items. Records are the horizontal components in a table. In SQL, they are called rows. The intersection of a field and a record is a single, discrete entity called a field value, or item value. In SQL, the intersection is called a datum.

A table in ArcView references tabular data from several sources--dBASE, INFO or delimited text files--in a uniform display format composed of fields and records. ArcView tables are dynamic, in that they reference the source data rather than contain the data itself. Consequently, a table in ArcView represents the current state of the data at the time the project is opened, including all changes in the data occurring subsequent to the last time the project was accessed.

Tables can be displayed, queried and analyzed. In addition, tables joined to a theme can be queried and analyzed spatially. Tables can also be joined based on equivalent values of a common field, even if the underlying format for the source data differs (such as dBASE and INFO).

Thematic Classification

Thematic classification refers to the assignment of symbols to features in a theme based on values for a field in the theme's attribute table, including tables which have been joined to the theme attribute table. A classification may be applied to a theme based on a range of values or unique values for the field. In ArcView, the maximum number of classes, regardless of the classification type, is 64.

Thematic Map

A thematic map is a map displaying a set of related geographic features. Typically, a classification has been applied so that the map displays attribute information associated with the geographic features. Examples are maps displaying land use or soil drainage classes.

Theme

A theme is a spatial data set (geography) linked with attribute data which contain a locational component. Imagine a map of a census tract, upon which you superimpose two other maps. The other two maps represent streets and a pizza chain's stores. Each of these maps would be a separate theme.

A theme is a group of similar geographic features in a view. A theme is based on a set of features of a specific feature class from a spatial data source. Feature classes include regions, routes, polygons, arcs, points, annotation, labels, and nodes. Certain feature classes, such as regions or routes, are complex classes made up of more basic feature types. The three basic feature types are points, lines (arcs) and polygons.

A theme can be comprised of all features within a specific feature class from a spatial data source, such as all land use polygons, or a subset of features from a spatial data source, such as all commercial land use polygons.

A spatial data source may contain more than one feature class. For example, a census geography may contain both lines which represent streets, and polygons which represent census tracts.

Vector Data

Vector data sets, also referred to as arc-node data, contain features with discrete positions. These positions are stored as x,y coordinate pairs. Vector data consist of a series of nodes that define line segments, which are in turned joined to form more complex features, such as line networks and polygons. ARC/INFO coverages and ArcView shape files are both examples of vector data. (See also: *Line Features, Polygon Features.*)

Vertex

A vertex is a specific x,y coordinate pair which makes up a line. Vertices are often referred to as shape points. The more vertices or shape points making up the line, the more accurate the representation of the feature. (See also: *Line Features.*)

View

Essentially, a view is a collection of themes. Assume you are examining the market for home improvement products in a three-county area in the state of New Mexico (Bernalillo, Sandoval and Valencia). You want geographic detail, including streets and roads, and zip code areas. You also want lifestyle characteristics of the people who live in specific locales and neighborhoods in the tri-county area. One of your views could consist of four themes: a map of the tri-county area, the street net, zip code areas, and a market segmentation overlay which defines market segments by zip code area. You could choose to display all themes in the view simultaneously, with the tri-county land map on the bottom, and the market segment theme on top. You could also choose to display only two themes in the view simultaneously, and so on.

In ArcView, views organize themes. A view window contains a graphics display area and a Table of Contents which lists all themes present in the view. You select which theme or themes to draw in the view.

More specifically, a view is an interactive map, in that all themes present in a view share a common geographic coordinate system, and most commonly share a coincident geographic extent (area) as well.

The view is the primary document type in ArcView. Other ArcView documents (tables, charts, and layouts) are typically linked to themes contained in views. A view imposes an organizational structure on themes derived from spatial data, and provides a means to display, query and analyze the data. This is accomplished by the interaction of the three primary components of a view--the map display, the table of contents, and the view graphical user interface. (See also: *Theme*.)

X,Y Coordinates

X,Y coordinates refer to the unique geographic location of a spatial feature. All spatial data is maintained in some form of map coordinate system, the most widely used being the geographic coordinates of latitude and longitude.

X,Y Events

X,Y events are exact locations of features on a map using x,y coordinates. Commonly used map coordinate systems include latitude/longitude, UTM, and State Plane Coordinates. (See also: *Event Table*.)

Appendix A
Installation and Configuration

It is not our purpose to provide step-by-step instructions for installing ArcView. Installation is straightforward, and the ArcView Installation Guide provides comprehensive instructions. Our purpose here is offer a few general observations that can improve ArcView's stability and performance.

Memory Considerations

Memory constraints have been a major issue throughout the development of ArcView version 2. Developers have struggled to accommodate the demand for increased functionality while at the same time retaining the ability to run ArcView on a "standard" office PC. At the time version 2 was released, a standard office PC could be defined as a PC with a 486-66 processor and 8 MB of RAM. The final release achieved this goal, but a question remains: is running ArcView on an 8 MB machine practical for your purposes?

Based on our personal observations and contacts with several ArcView user sites, while ArcView will run on an 8 MB machine, you may not be happy with performance. If this is the case, then the money to upgrade to 16 MB would be well spent. ESRI requires a minimum of 12 MB swap space under Windows on your hard disk in addition to 8 MB of RAM. ArcView spends a lot of time swapping to disk on an 8 MB machine which degrades overall performance. Users have reported that swapping is substantially reduced with 16 MB of RAM, and performance is entirely acceptable.

Memory management during an ArcView session is another matter. During ArcView operation, a certain amount of memory is associated with each object. As you work in a project, the total amount of memory utilized increases as you open and manipulate different documents. As you switch operations and documents, ArcView flags the memory associated with the old objects as available for reuse, but does *not* immediately free up this memory. This memory is reallocated only when ArcView needs to load new objects. The result is that while your ArcView session will indefinitely run in a stable manner, you may have a problem maintaining enough memory to run additional applications. This may become evident when initializing an additional operation with a high memory requirement, such as printing a view or layout. It may be necessary to close additional applications before printing, or optionally, to print to a file so that the document can be sent to the printer after the application has been closed.

A word of caution: Because ArcView associates a certain amount of memory with each object, operations involving repeated manipulation of many objects (e.g., editing shape files or manipulating chart markers) may eventually consume all available memory if continued over an extended period of time. Our advice is to listen for hard disk activity. If you begin to hear a significant amount of disk activity during an intensive editing session, this likely indicates that ArcView is having to swap large amounts of data from memory to disk. Stop and wait for the system to catch up, then save your work. Closing and reopening the ArcView document may also serve to reallocate memory.

Configuration Considerations

In the course of developing this book the ArcView environment was found to be very stable. However, some sites have reported intermittent troubles. We have noted a common thread among many of these problems, and have discovered a solution which addresses the problems.

ArcView is a 32-bit application. In order to run ArcView under Windows, an extension to Windows allowing 32-bit applications to run must be present. The ArcView installation process installs these extensions if they are not already present on your system. However, ArcView does *not* write over 32-bit extensions to Windows if they are already present on your system.

If you installed a pre-release version of ArcView, an early version of these 32-bit extensions was installed at that time. If you subsequently installed the final version of ArcView, the 32-bit Windows extensions were *not* updated with the later versions supplied with the final release. This can result in intermittent segmentation violations, system crashes, and other problems while running ArcView. If you have installed the final version of ArcView on top of the pre-release version, or had the 32-bit extensions to ArcView already installed from another application, you should delete the 32-bit Windows extensions, as described on page 22 of the ArcView Installation Guide, and perform a *clean* installation.

Operating Systems

PC operating systems are in a transitional state. While Windows 3.1/Windows for Workgroups 3.11 is the standard environment in which ArcView is being run at present, alternatives do exist. Many users have asked whether performance gains might be obtained by switching to Windows NT or OS/2.

Both Windows NT and OS/2 are 32-bit operating systems. As ArcView 2.0 is a 32-bit application, intuitively it makes sense to run ArcView in such an environment. ArcView 2.0 will run under Windows NT 3.5 and OS/2 Warp 3.0. While no quantitative comparisons are yet available as to how well ArcView runs on either system, the initial word from users is that you can expect a performance increase under Windows NT over Windows 3.1.

However, these potential performance increases come at the expense of system resources. OS/2 Warp requires 35 to 65 MB of disk space for installation depending on configuration, while Windows NT requires 70 MB of disk space. Windows NT also requires a minimum of 12 MB of RAM, with 16 MB desirable. In addition, OS/2 Warp 3.0 requires that Windows be installed in order to run Windows applications under OS/2.

Appendix B

Functionality Quick Reference

This appendix is a reference to the functionality available through the ArcView menu, button and tool bars. It is designed to be a quick reference to the main functionality available in ArcView, as well as to where these topics are covered in the book.

Add Event Theme
Access: View menu bar—View menu
Purpose: Adds a new theme to a view using an event table. An event table contains a locatable field such as an address, XY coordinates, or a route location.
Associated with: Add Theme, Event Tables, Address Geocoding
Reference: Chapters 2, 4

Add Field
Access: Table menu bar—Edit menu
Purpose: Adds a new field to the active table. The table must be in dBASE or INFO format.
Associated with: Start/Stop Editing, Add Record, Delete Field
Reference: Chapter 9

Add Record
Access: Table menu bar—Edit menu

282   Appendix B: Functionality Quick Reference

Purpose: Adds a new record to the active table which has been opened to allow editing.
Associated with: Start/Stop Editing, Add Field, Delete Record
Reference: Chapter 9

Add Table
Access: Project menu bar—Project menu
Purpose: Imports a dBASE, INFO or delimited text file into the active project.
Associated with: Import Project, SQL Connect
Reference: Chapter 3

Add Theme
Access: View menu bar—View menu; View button bar
Purpose: Adds a theme to a view from a spatial data source, such as an ArcView shape file, ARC/INFO coverage, or supported image source.
Associated with: Delete Themes, New Theme
Reference: Chapter 3

Align
Access: View menu bar—Graphics menu; Layout menu bar—Graphics menu
Purpose: Aligns selected graphics vertically or horizontally in a view or layout.
Associated with: Pointer
Reference: Chapter 8

Area Chart Gallery
Access: Chart menu bar—Gallery menu; Chart button bar
Purpose: Displays format options and allows change of active chart to Area chart format.
Associated with: Create Chart, Bar Chart Gallery, Column Chart Gallery, Line Chart Gallery, Pie Chart Gallery, XY Scatter Chart Gallery
Reference: Chapter 7

Attach Graphics
Access: View menu bar—Theme menu
Purpose: Attaches the selected graphics to the active theme. Attached graphics will display only when the theme is turned on.
Associated with: Detach Graphics, Label Features, AutoLabel
Reference: Chapter 5

AutoLabel
Access: View menu bar—Theme menu
Purpose: Labels the selected features in the active themes using the specified label field. The resultant text graphics are attached to the active themes.
Associated with: Label Features, Text
Reference: Chapter 5

Bar Chart Gallery
Access: Chart menu bar—Gallery menu; Chart button bar
Purpose: Displays format options and allows change of active chart to Bar chart format.
Associated with: Create Chart, Area Chart Gallery, Column Chart Gallery, Line Chart Gallery, Pie Chart Gallery, XY Scatter Chart Gallery
Reference: Chapter 7

Bring To Front
Access: View menu bar—Graphics menu; Layout menu bar—Graphics menu; Layout button bar
Purpose: Brings selected graphics to the front of remaining graphics.
Associated with: Send To Back, Pointer
Reference: Chapter 8

Calculate
Access: Table menu bar—Field menu; Table button bar
Purpose: Performs calculations on all or selected records on the active field of a table for which editing has been enabled.
Associated with: Start/Stop Editing
Reference: Chapter 9

284 Appendix B: Functionality Quick Reference

Chart Color
Access: Chart tool bar
Purpose: Changes the color of any chart element.
Associated with: Show Symbol Palette
Reference: Chapter 7

Chart Element Properties
Access: Chart tool bar
Purpose: Sets properties of the chart elements of the active chart.
Associated with: Chart Axis Properties, Chart Legend Properties, Chart Title Properties
Reference: Chapter 7

Chart Properties
Access: Chart menu bar—Chart menu; Chart button bar
Purpose: Sets the properties of the active chart, including add/delete data series or groups, select fields for labeling data series/groups, and set order of displaying data series/groups.
Associated with: Create Chart, Table Properties
Reference: Chapter 7

Clear Selected Features
Access: View menu bar—Theme menu; View button bar
Purpose: Deselects any selected features in the active themes.
Associated with: Select Features, Select Features Using Shape, Switch Selection
Reference: Chapter 6

Close
Access: File menu—View, Table, Chart and Layout menu bars
Purpose: Closes the active component of a project.
Associated with: Close All

Close All
Access: File menu—View, Table, Chart and Layout menu bars
Purpose: Closes all project components currently open.
Associated with: Close

Close Project
Access: Project menu bar—File menu
Purpose: Closes the active project and all components.
Associated with: Close, Close All, Exit

Column Chart Gallery
Access: Chart menu bar—Gallery menu; Chart button bar
Purpose: Displays format options and allows change of active chart to Column chart format.
Associated with: Create Chart, Area Chart Gallery, Bar Chart Gallery, Line Chart Gallery, Pie Chart Gallery, XY Scatter Chart Gallery
Reference: Chapter 7

Convert to Shapefile
Access: View menu bar—Theme menu
Purpose: Converts a theme derived from an ARC/INFO coverage to ArcView shape file format for the entire theme or a selected set of features.
Associated with: New Theme, Add Theme
Reference: Chapter 9

Copy
Access: Table menu bar—Edit menu; Layout menu bar—Edit menu; Table and Layout button bar
Purpose: Copies the selected features to the clipboard. For tables, the selected feature is data in the active cell, and for layouts, graphics.
Associated with: Cut, Paste

Copy Graphics
Access: View menu bar—Edit menu
Purpose: Copies the selected graphics in a view to the clipboard.
Associated with: Cut, Paste

Copy Themes
Access: View menu bar—Edit menu
Purpose: Copies the active themes to the clipboard.
Associated with: Cut, Paste
Reference: Chapter 4

Create Chart
Access: Table menu bar—Table menu; Table button bar
Purpose: Creates a chart from the selected records of the active table, or the entire table if no records are selected.
Associated with: Chart Properties
Reference: Chapter 7

Create/Remove Index
Access: Table menu bar—Field menu
Purpose: Creates or removes an ArcView index for the active field. If the active field is the Shape field, a spatial index will be created or deleted.
Associated with: Open Theme Table
Reference: Chapter 10

Customize
Access: Project menu bar—Project menu
Purpose: Accesses the dialog box for customizing the ArcView interface. This selection is accessible only if Avenue has been installed.
Associated with: Control Properties, Project Properties
Reference: Chapter 11

Cut
Access: Table menu bar—Edit menu; Layout menu bar—Edit menu; Table button bar; Layout button bar
Purpose: Cuts the selection and places it on the clipboard. For tables, the selection would be data in the active cell, and for layouts, graphics.
Associated with: Copy, Paste

Cut Graphics
Access: View menu bar—Edit menu
Purpose: Cuts the selected graphics from the view and places them on the clipboard.
Associated with: Cut, Copy, Paste, Delete Graphics

Cut Themes
Access: View menu bar—Edit menu
Purpose: Cuts the active themes from the view and places them on the clipboard.
Associated with: Copy Themes, Paste, Delete Themes

Delete
Access: Layout menu bar—Edit menu
Purpose: Deletes the selected graphics in the active layout.
Associated with: Delete Graphics, Cut, Cut Graphics

Delete Field
Access: Table menu bar—Edit menu
Purpose: Deletes the active field from a table in which editing has been enabled.
Associated with: Add Field, Start/Stop Editing
Reference: Chapter 9

288 *Appendix B: Functionality Quick Reference*

Delete Graphics
Access: View menu bar—Graphics menu
Purpose: Deletes selected graphics from a view.
Associated with: Cut Graphics, Copy Graphics

Delete Records
Access: Table menu bar—Edit menu
Purpose: Deletes the selected records for the active table, providing editing has been enabled.
Associated with: Cut, Copy, Paste, Add Records
Reference: Chapter 9

Delete Themes
Access: View menu bar—Edit menu
Purpose: Deletes the active themes from a view.
Associated with: Cut Themes, Copy Themes

Detach Graphics
Access: View menu bar—Theme menu
Purpose: Detaches the graphics from the active themes in the view.
Associated with: Attach Graphics
Reference: Chapter 5

Draw
Access: View tool bar; Layout tool bar
Purpose: Allows graphics (points, lines, polylines, rectangles, circles and polygons) to be added to a view or layout.
Associated with: Text, Label, Pointer
Reference: Chapter 5

Edit
Access: Table tool bar
Purpose: Selects cells for editing data in an active table for which editing has been enabled.

Associated with: Start/Stop Editing
Reference: Chapter 9

 ## Edit Legend
Access: View menu bar—Theme menu; View button bar
Purpose: Accesses the Legend Editor for changing the symbols and/or classification of the active theme.
Associated with: Theme Properties
Reference: Chapter 5

 ## Erase
Access: Chart tool bar
Purpose: Removes data markers from a chart and unselects the records from the associated table.
Associated with: Erase With Polygon, Undo Erase

 ## Erase With Polygon
Access: Chart tool bar
Purpose: Removes one or more data markers from an XY Scatter chart by defining a polygon around the markers.
Associated with: Erase, Undo Erase

Exit
Access: File menu—Project, View, Table, Chart and Layout menu bars
Purpose: Ends the ArcView session.
Associated with: Close Project, Save Project, Save Project As

Export
Access: View menu bar—File menu; Layout menu bar—File menu
Purpose: Exports a view or layout to a file. Supported output formats include Encapsulated PostScript, Adobe Illustrator, and CGM on all platforms, as well as Windows Metafile and Bitmap, and Macintosh PICT.
Associated with: Print, Print Setup

290 Appendix B: Functionality Quick Reference

Export Table
Access: Table menu bar—File menu
Purpose: Exports a table to a file. Supported file types include dBASE, INFO and delimited text.
Associated with: Print, Print Setup
Reference: Chapter 9

Find
Access: View menu bar—View menu; Table menu bar—Table menu; Chart menu bar—Chart menu; View, Table, and Chart button bars
Purpose: Finds a particular feature in the active theme, table, or chart based on an entered text string.
Associated with: Locate, Query
Reference: Chapter 6

Frame
Access: Layout tool bar
Purpose: Adds a view frame, legend frame, scale bar frame, north arrow frame, chart frame, table frame, or picture frame to a layout.
Associated with: Draw, Text
Reference: Chapter 8

Group
Access: View menu bar—Graphics menu; Layout menu bar—Graphics menu; Layout button bar
Purpose: Groups selected graphics into a single graphic.
Associated with: Ungroup, Pointer
Reference: Chapter 8

Hide/Show Legend
Access: View menu bar—Theme menu
Purpose: Hides or shows the legend of the active themes in the Table of Contents.
Associated with: Edit Legend

Hot Link
Access: View tool bar
Purpose: Invokes a defined hot link for the active theme on a view.
Associated with: Open View, Open Project
Reference: Chapter 9

 ## Identify
Access: View, Table, and Chart tool bars
Purpose: Displays the attributes of a feature in an active theme, table or chart.
Associated with: Select Feature, Select, Select Features Using Shape
Reference: Chapter 5

Import
Access: Project menu bar—Project menu
Purpose: Imports the components of another ArcView project, including ArcView 1.0 projects, into the active project.
Associated with: Open Project

 ## Join
Access: Table menu bar—Table menu; Table button bar
Purpose: Joins a table to the active table based on the values of a common field.
Associated with: Link, Remove All Joins
Reference: Chapters 4, 9

 ## Label
Access: View tool bar
Purpose: Labels a feature in the active theme using the attribute from the field specified in the theme's properties.
Associated with: AutoLabel, Text
Reference: Chapter 5

Layout
Access: View menu bar—View menu
Purpose: Creates a layout using a specified template.
Associated with: Use Template, Store as Template
Reference: Chapter 8

Line Chart Gallery
Access: Chart menu bar—Gallery menu; Chart button bar
Purpose: Displays format options and allows change of active chart to Line chart format.
Associated with: Create Chart, Area Chart Gallery, Bar Chart Gallery, Column Chart Gallery, Pie Chart Gallery, XY Scatter Chart Gallery
Reference: Chapter 7

Link
Access: Table menu bar—Table menu
Purpose: Establishes a one-to-many relationship between the source table to the active table based on the values of a common field.
Associated with: Join, Remove All Links
Reference: Chapter 4

Locate
Access: View menu bar—View menu; View button bar
Purpose: Locates a specific address on an active theme for which geocoding properties have been set.
Associated with: Find
Reference: Chapter 4

Measure
Access: View tool bar
Purpose: Measures distance on a view.
Associated with: Draw

Merge Graphics
Access: View menu bar—Edit menu
Purpose: Combines or aggregates selected features in an ArcView shape file into a single shape.
Associated with: Select Feature, Summarize
Reference: Chapter 9

New Project
Access: Project menu bar—File menu
Purpose: Creates a new ArcView project.
Associated with: Save Project As
Reference: Chapter 3

New Theme
Access: View menu bar—View menu
Purpose: Creates a new theme in a view based on the ArcView shape file format.
Associated with: Edit, Copy Theme
Reference: Chapter 9

Open Project
Access: Project menu bar—File menu
Purpose: Opens an existing ArcView project.
Associated with: New Project, Save Project, Save Project As
Reference: Chapter 3

Open Theme Table
Access: View menu bar—Theme menu; View button bar
Purpose: Opens the attribute tables for the active themes in a view.
Associated with: Add Table
Reference: Chapter 6

Page Setup
Access: Layout menu bar—Layout menu

Purpose: Defines the characteristics of the layout page.
Associated with: Layout Properties, Use Template, Show/Hide Grid, Show/ Hide Margins
Reference: Chapter 8

Pan
Access: View tool bar; Layout tool bar
Purpose: Pans the view or layout by dragging the display with the mouse.
Associated with: Zoom In, Zoom Out

Paste
Access: Edit menu—View, Table and Layout menu bars; Table button bar; Layout button bar
Purpose: Pastes the contents of the clipboard into the active document.
Associated with: Copy, Cut

Pie Chart Gallery
Access: Chart menu bar—Gallery menu; Chart button bar
Purpose: Displays format options and allows change of active charts to Pie chart format.
Associated with: Create Chart, Area Chart Gallery, Bar Chart Gallery, Column Chart Gallery, Line Chart Gallery, XY Scatter Chart Gallery
Reference: Chapter 7

Pointer
Access: View tool bar; Layout tool bar
Purpose: Selects graphics in a view or layout for subsequent editing and manipulation.
Associated with: Select All, Select All Graphics
Reference: Chapter 5

Print
Access: File menu—View, Table, Chart and Layout menu bars; Layout button bar

Purpose: Prints the active project component.
Associated with: Print Setup

Print Setup
Access: File menu—View, Table, Chart and Layout menu bars
Purpose: Controls the output format and printing environment.
Associated with: Print

Promote
Access: Table menu bar—Table menu; Table button bar
Purpose: Displays selected records at the top of the table.
Associated with: Select, Sort Ascending, Sort Descending
Reference: Chapter 6

Properties—Graphic
Access: View menu bar—Graphics menu; Layout menu bar—Graphics menu
Purpose: Displays and edits graphics properties for graphic primitives, text and frames.
Associated with: View Properties, Layout Properties, Pointer, Text
Reference: Chapter 8

Properties—Layout
Access: Layout menu bar—Layout menu; Layout button bar
Purpose: Display and edit layout properties, including name, grid spacing, and snap to grid.
Associated with: Graphics Properties, View Properties, Table Properties, Chart Properties
Reference: Chapter 8

Properties—Project
Access: Project menu bar—Project menu
Purpose: Display and edit project properties, including start-up and shutdown scripts, work directory, name of creator and creation date, and selection color.

296 Appendix B: Functionality Quick Reference

Associated with: View Properties, Table Properties, Chart Properties, Layout Properties
Reference: Chapter 9

Properties—Table

Access: Table menu bar—Table menu
Purpose: Display and edit properties of the active table, including name, creator, visible fields, and field alias names.
Associated with: View Properties, Chart Properties, Layout Properties
Reference: Chapter 6

Properties—View

Access: View menu bar—View menu
Purpose: Display and edit the properties of the current view, including name, creation date, creator, map units, distance units, and projection.
Associated with: Table Properties, Chart Properties, Layout Properties
Reference: Chapter 3

Query

Access: View menu bar—Theme menu; Table menu bar—Table menu; View button bar; Table button bar
Purpose: Opens the Query Builder dialog window which allows the feature(s) in a view or records in a table to be selected by a logical expression based on attribute values.
Associated with: Find, Locate, Theme Properties
Reference: Chapter 6

Refresh

Access: Table menu bar—Table menu
Purpose: Causes ArcView to re-read the source data for the active table.
Associated with: Open, Add Table, Join, Link
Reference: Chapter 7

Rematch

Access: View menu bar—Theme menu

Purpose: Opens the Geocoding Editor dialog window, allowing features to be rematched in the geocoded theme.
Associated with: Add Event Theme
Reference: Chapter 4

Remove All Joins
Access: Table menu bar—Table menu
Purpose: Removes all joins on the active table.
Associated with: Join, Remove All Links
Reference: Chapter 4

Remove All Links
Access: Table menu bar—Table menu
Purpose: Removes all links to other tables for the active table.
Associated with: Link, Remove All Joins
Reference: Chapter 4

Rename
Access: Project menu bar—Project menu
Purpose: Renames the selected project component.
Associated with: Properties

Save Project
Access: File menu—Project, View, Table, Chart and Layout menu bars; Project, View, Table, Chart and Layout button bars
Purpose: Saves the active project.
Associated with: Save Project As
Reference: Chapter 3

Save Project As
Access: Project menu bar—File menu
Purpose: Saves the active project to a new name and/or directory.
Associated with: Save Project
Reference: Chapter 3

298   Appendix B: Functionality Quick Reference

Select
Access: Table tool bar
Purpose: Selects records in the active table.
Associated with: Select All, Select None, Switch Selection
Reference: Chapter 6

Select All
Access: Table menu bar—Edit menu; Layout menu bar—Edit menu; Table button bar
Purpose: Selects all records in the active table or all graphics drawn in the active layout.
Associated with: Pointer, Select, Select Name, Switch Selection
Reference: Chapter 6

Select All Graphics
Access: View menu bar—Edit menu
Purpose: Selects all graphics drawn in the view.
Associated with: Pointer

Select By Theme
Access: View menu bar—Theme menu
Purpose: Selects features of the active themes based on selected features of the selector theme.
Associated with: Select Feature, Select Features Using Shape
Reference: Chapter 9

Select Feature
Access: View tool bar
Purpose: Selects features in the active theme using the mouse.
Associated with: Select Features Using Shape, Select By Theme
Reference: Chapter 6

Select Features Using Shape
Access: View button bar

Purpose: Selects features in the active theme using selected graphics in the view.
Associated with: Select Feature, Select By Theme, Pointer
Reference: Chapter 6

Select None
Access: Table menu bar—Edit menu; Table button bar
Purpose: Clears the selected set in the active table.
Associated with: Select, Select All, Switch Selected
Reference: Chapter 6

Send to Back
Access: View menu bar—Graphics menu; Layout menu bar—Graphics menu; Layout button bar
Purpose: Places the selected graphics behind the remaining graphics.
Associated with: Bring to Front, Pointer
Reference: Chapter 8

Series From Records/Fields
Access: Chart menu bar—Chart menu; Chart button bar
Purpose: Toggles the plotting of the data series in a chart from records or fields.
Associated with: Create Chart, Chart Properties
Reference: Chapter 7

Show/Hide Grid
Access: Layout menu bar—Layout menu
Purpose: Toggles the display of the layout grid on the active layout.
Associated with: Layout Properties, Show/Hide Margins
Reference: Chapter 8

Show/Hide Legend
Access: Chart menu bar—Chart menu
Purpose: Toggles the display of the legend on the active chart.

Associated with: Chart Properties, Show/Hide X Axis, Show/Hide Y Axis, Show/Hide Title

Show/Hide Margins
Access: Layout menu bar—Layout menu
Purpose: Toggles the display of the layout page margins on the active layout.
Associated with: Layout Properties, Show/Hide Grid
Reference: Chapter 8

Show/Hide Title
Access: Chart menu bar—Chart menu
Purpose: Toggles the display of the title on the active chart.
Associated with: Chart Properties, Show/Hide Legend, Show/Hide X Axis, Show/Hide Y Axis
Reference: Chapter 7

Show/Hide X Axis
Access: Chart menu bar—Chart menu
Purpose: Toggles the display of the X Axis with tick marks on the active chart.
Associated with: Chart Properties, Show/Hide Y Axis, Show/Hide Legend, Show/Hide Title
Reference: Chapter 7

Show/Hide Y Axis
Access: Chart menu bar—Chart menu
Purpose: Toggles the display of the Y Axis with tick marks on the active chart.
Associated with: Chart Properties, Show/Hide X Axis, Show/Hide Legend, Show/Hide Title
Reference: Chapter 7

Show Symbol Palette
Access: Window menu—View, Table, Chart and Layout menu bars

Purpose: Displays the symbol palette.
Associated with: Legend Editor

Size and Position

Access: View menu bar—Graphics menu; Layout menu bar—Graphics menu
Purpose: Displays the dialog box for controlling the size and position of the selected graphics.
Associated with: Graphics Properties, Pointer
Reference: Chapter 8

Snap

Access: View tool bar
Purpose: Sets the tolerance for snapping vertices on an editable theme for which snapping has been enabled.
Associated with: Start/Stop Editing, Draw, Pointer
Reference: Chapter 9

Sort Ascending / Sort Descending

Access: Table menu bar—Field menu; Table button bar
Purpose: Sorts all records in the active table on the active field.
Associated with: Promote
Reference: Chapter 6

SQL Connect

Access: Project menu bar—Project menu
Purpose: Opens the SQL Connect dialog window to enable a connection to a database server and subsequent retrieval of records based on an SQL query.
Associated with: Export

Start/Stop Editing

Access: Table menu bar—Table menu; View menu bar—Theme menu
Purpose: Controls whether editing is enabled on a theme or table.

Associated with: Edit, Pointer
Reference: Chapter 9

Statistics
Access: Table menu bar—Field menu
Purpose: Obtains statistics about an active numeric field in the active table.
Associated with: Summarize, Query
Reference: Chapter 9

Store As Template
Access: Layout menu bar—Layout menu
Purpose: Creates a layout template from the current layout.
Associated with: Use Template
Reference: Chapter 8

Summarize
Access: Table menu bar—Field menu; Table button bar
Purpose: Displays the Summary Table Definition dialog box to prepare a summary table based on the active field.
Associated with: Statistics, Merge, Query
Reference: Chapter 9

Switch Selection
Access: Table menu bar—Edit menu; Table button bar
Purpose: Switches the selected set of records in the active table to all records previously unselected.
Associated with: Select, Select All, Select None
Reference: Chapter 6

Text
Access: View tool bar; Layout tool bar
Purpose: Adds or edits text in the active view or layout.
Associated with: Pointer, Draw
Reference: Chapter 5

Theme Properties
Access: View menu bar—Theme menu; View button bar
Purpose: Review and set the properties of the active theme, including name, logical queries, field for feature labeling, range of scales for display, hot link definition, geocoding properties, and snapping.
Associated with: View Properties, Edit Legend
Reference: Chapter 3

Themes On / Themes Off
Access: View menu bar—View menu
Purpose: Turns all themes in a view on or off.
Associated with: Theme Properties
Reference: Chapter 3

 ### Undo Erase
Access: Chart menu bar—Edit menu; Chart button bar
Purpose: Undeletes the last data markers erased from the active chart.
Associated with: Erase, Erase With Polygon

 ### Ungroup
Access: View menu bar—Graphics menu; Layout menu bar—Graphics menu
Purpose: Ungroups a selected previously grouped graphic into the original individual graphics.
Associated with: Group, Pointer
Reference: Chapter 8

Use Template
Access: Layout menu bar—Layout menu; View menu bar—View menu
Purpose: Creates a new layout or updates the current layout using a specified stored template.
Associated with: Store As Template, Layout Properties
Reference: Chapter 8

XY Scatter Chart Gallery
Access: Chart menu bar—Gallery menu; Chart button bar
Purpose: Displays format options and allows change of active chart to XY Scatter chart format.
Associated with: Create Chart, Area Chart Gallery, Bar Chart Gallery, Column Chart Gallery, Line Chart Gallery
Reference: Chapter 7

 ### Zoom In
Access: View menu bar—View menu; Layout menu bar—Layout menu; View button bar; Layout button bar
Purpose: Zooms into the center of the active view or layout by a factor of 2.
Associated with: Zoom Out, Zoom to Full Extent, Zoom to Selected, Zoom to Themes, Zoom to Page, Zoom to Actual Size

 ### Zoom In (Tool)
Access: View tool bar; Layout tool bar
Purpose: Zooms in at the position you click or the area you describe on a view or layout.
Associated with: Zoom Out, Zoom to Full Extent, Zoom to Selected, Zoom to Themes, Zoom to Page, Zoom to Actual Size

 ### Zoom Out
Access: View menu bar—View menu; Layout menu bar—Layout menu; View button bar; Layout button bar
Purpose: Zooms out from the center of the active view of layout by a factor of 2.
Associated with: Zoom In, Zoom to Full Extent, Zoom to Selected, Zoom to Themes, Zoom to Page, Zoom to Actual Size

 ### Zoom Out (Tool)
Access: View tool bar; Layout tool bar
Purpose: Zooms out at the position you click or the area you describe on a view or layout.

Associated with: Zoom In, Zoom to Full Extent, Zoom to Selected, Zoom to Themes, Zoom to Page, Zoom to Actual Size

Zoom to Actual Size
Access: Layout menu bar—Layout menu; Layout button bar
Purpose: Zooms to the actual (1:1) size of the layout page.
Associated with: Zoom In, Zoom Out, Zoom to Page, Zoom to Selected
Reference: Chapter 8

Zoom to Full Extent
Access: View menu bar—View menu; View button bar
Purpose: Zooms to the full extent of all themes in a view.
Associated with: Zoom In, Zoom Out, Zoom to Selected, Zoom to Themes
Reference: Chapter 3

Zoom to Page
Access: Layout menu bar—Layout menu; Layout button bar
Purpose: Zooms to the full extent of the layout page.
Associated with: Zoom In, Zoom Out, Zoom to Actual Size, Zoom to Selected
Reference: Chapter 8

Zoom to Selected
Access: View menu bar—View menu; Layout menu bar Layout menu; View button bar; Layout button bar
Purpose: Zooms to the selected features of the active themes in a view, or to the selected graphics of a layout.
Associated with: Zoom In, Zoom Out, Zoom to Full Extent, Zoom to Themes, Zoom to Page, Zoom to Actual Size
Reference: Chapter 6

Zoom to Themes
Access: View menu bar—View menu; View button bar
Purpose: Zooms to the extent of the active themes in a view.

Associated with: Zoom In, Zoom Out, Zoom to Full Extent, Zoom to Selected
Reference: Chapter 3

Appendix C
About the MicroVision Segments

In the book, we referred to several lifestyle segmentation names marketed by Equifax National Decision Systems. Appearing below are descriptions of the nine primary Tempe segments for block groups associated with survey responses.

Note that for purposes of demonstration, we elected to use the general purpose MicroVision segments. A total of 50 general purpose segments exist. Specific, industry-tailored versions of these segments are also available from Equifax. In our restaurant exercises, for example, we could have easily employed the MicroVision Restaurant data set.

The Primary MicroVision Segments for Tempe

Prosperous Metro Mix. Only 2% of the population is located in rural areas. Ranking fourth (of 50 segments) in average household size, this segment also ranks high on the number of adults in the 30 to 39 age cohort, and above average for children age 0 to 17. This segment earns the fifth highest median household income at 75% above the national average. Over 70% of the families include two or more workers. This segment ranks high in the number of working female adults with children, including the highest number of working mothers with children under age six.

Movers and Shakers. This highly urban segment ranks second in the total number of two-person households (37%). Median household income is 55% higher than the national average, and the segment ranks fifth in per capita income. The segment ranks fourth in the number of adults with undergraduate and graduate degrees, fifth in total white collar employment (81%) and third in the number of workers in professional specialties (27%).

Home Sweet Home. With median household income 42% higher than the national average, this segment ranks third among the 50 segments in the proportion of households earning between $35,000 and $50,000. Nearly eight of ten households own their home. This segment ranks fourth in the proportion of properties valued between $100,000 and $150,000.

A Good Step Forward. Only 1% of the population resides in rural areas. This segment ranks among the top four for the number of individuals age 22 to 34. Along with an above average senior population (65 and older), this segment ranks among the lowest for children age 0 to 17. This segment ranks fifth in the number of persons in non-family households (33%), and fourth in the percentage of single-person households (43%).

Great Beginnings. Ranking at the top of the 50 segments for adults with either some college or an associate degree, this segment is also above average for higher education degrees attained. Less than half of the segment are living in their own homes, at 25% below the national average. Median property value is high and may contribute to rent payments which are nearly 40% above the national norm.

White Picket Fences. A segment of average age, the population is evenly distributed among the age ranges in accordance with the national average. Household size is also average, as is the proportion of married individuals and family households. Ranking fifth on the proportion of adults with only a high school diploma, this segment is well below average on the proportion of graduate and undergraduate degrees, and ranks second on the proportion of households earning between $25,000 and $35,000 (19%).

Books and New Recruits. With 40% of all individuals between ages 18 and 24, this segment is the lowest ranking segment for median age and near the bottom for the proportion of married and divorced individuals and family households. Over 28% of the population resides in group quarters such as dormitories and barracks, and 67% of students are enrolled in college. This segment has the shortest average commute time, with over 35% of its workers taking less than 10 minutes to travel to work.

University USA. This is the number one segment for persons living in group quarters (30%), and ranks first for the proportion of non-family households (26%). Only one in ten persons are under age 18. Of all persons identified as students, 89% are presently attending college. Over 75% of housing is renter-occupied, the fourth highest of any segment.

Urban Singles. With a median age 23% higher than average, this segment has concentrations of individuals age 18 to 29 and of persons age 70 and older. This is the number one segment for adults over the age of 84 and also ranks first in the proportion of single-person households. Nearly one-third of the adults do not have a high school diploma. Employment is noticeably high in the service occupations. An above average proportion of workers use public transportation or walk to work. This segment ranks fifth in the proportion of households without vehicles (38%).

Index

A

Academic use of ArcView
   University of Missouri at St. Louis 255
   University of Wisconsin-Madison 257
Add Event Theme 281
Add Event Themes
   Dialog box illustration 38
   Icon illustrated 36
Add Field 281
Add Record 281
Add Table 282
Add Table dialog window 52–53
Add Theme 282
Add Theme dialog window 52, 54
Add Theme icon, illustration 20
Adding data
   Add Table dialog window 52–53
   Add Theme dialog window 52
   Directories 53
   Libraries 53
   Project pull-down menu 53
   Project window, accessing data from 53
Addresses
   Address events 76
   Address matched event theme illustrated 41
   Site selection sample project 36
Alias, names for fields in tables 113
Align 282
Align tool 175
American Isuzu's use of ArcView 254
Application window, ArcView 14
ARC Macro Language, Avenue communication with 244
ARC/INFO
   ArcStorm Libraries 81
   ArcView's advantages xii
   CONVERTWORKSPACE command 58

Coverage, defined 20
Exporting data 57
GRID data 207
Importing data 57
Libraries 81
Maintenance of project, coverages 230
Map projections, transforming 71
Naming conventions 80
Shape files, converting ARC/INFO coverages to 227
Shape files, converting themes to 194
UNIX xii
ArcStorm Libraries 81
ArcView
   Application window 13, 15
   ARC/INFO access xii
   Avenue. *See* Avenue
   Buttons 14
   Converting spatial data xiii
   Customization 7
   Described 1–14
   Desktop mapping application described 1
   Desktop mapping application origins 2–3
   Environment and interface 13, 15
   Flexibility 7
   Functionality xii
   Icons 13, 15
   Integration 8
   Main window, illustration 13, 15
   Networks, accessibility by 8
   Object class hierarchy, Avenue script creation 242
   Power 7
   Project window 14
   Pull-down menus 14
   Sample project 18
   Status bar 14
   Table template 113

Themes, defined 20
Use with other platforms 8
Window elements, illustration 14
ArcView Windows xii
Are Completely Within, spatial join 192
Are Completely Within, theme-on-theme selection 189
Are Within Distance Of, theme-on-theme selection 190
Area Chart Gallery 282
Area chart, illustrated 147
Area in graphics, setting 103
Arguments, Avenue 238, 240
Arizona State Land Department's use of ArcView 249
ASCII files
 Importing data to ArcView 25
 Tabular formats supported by Arcview 50
Attach Graphics 283
Attributes
 Shape files, adding attributes to themes 198
 Shape files, geographic features 194
 Shape files, merging features by attribute value 200
 Table creation 161
Attributes of Tracts table
 Illustrated 28
 Joined table, illustration 30
Auto Label 283
Auto-Label tool 138
Automatic thematic maps 2
Avenue
 ArcView xii
 ArcView's advantages 8, xii
 Arguments 238, 240
 As object oriented language 236
 Class, common inherited properties 237
 Class, objects as instances of 237
 Classes of objects 237
 Communication with other applications, command strings 244

Communication with other applications, IAC 244
Communication with other applications, System.Execute 244
Hot links, script execution 193
Objects as instances of class 237
Properties of objects 237
Requests 240
Requests, object interaction by 238
Script creation, ArcView object class hierarchy 242
Script creation, converting with Compile button 241
Script creation, Customize Dialog window 242
Script creation, debugging 241
Script creation, Examine Variables button 242
Script creation, Help system 242
Script creation, linking with ArcView control 242
Script creation, Script window 241
Script creation, Search 242
Script creation, steps 240–241, 243
Scripts, defined 238
Scripts, document categories 15
Scripts, embedding 246
Scripts, encrypting 247
Scripts, locking projects 247
Scripts, syntax 240
Scripts, Zoom In 238
Start-up and shutdown scripts, customization 245
Start-up and shutdown scripts, global changes 246
Start-up and shutdown scripts, prompting the user 245
Statements 238
See also Graphical user interface customization

B

Background, moving graphics to   175
Bar Chart Gallery   283
Bar chart, illustrated   147
Box tool   175, 187
Bring To Front   283
Buttons
   Compile button, Avenue script conversion   241
   Delete button   234
   Described   14
   Examine Variables button, Avenue script creation   242
   Make Default button for GUI customization   236
   New button   234
   Query Builder   118
   Ramp   97
   Random   97
   Reset button, saving GUI customization   236
   Save button for GUI customization   236
   Separator button   235
   Sort Ascending   115
   Sort Descending   115
   Switch Selection   119
   Zoom In   24
   Zoom Out   24
   Zoom to Active Themes   24
   Zoom to Full Extent   24

C

Calculate   283
Cartographic design
   Defined   176
   Five Ws of communication   177
   Style development   178
Chart Color   284
Chart Element Properties   284
Chart exercise
   Attribute table creation   161
   Chart generation   152
   Field, adding to table   157
   Joining tables   152
   MicroVision segments, adding   152
   Series From Records/Series From Fields icon   155
   Summary table creation   156, 163
Chart Properties   284
Charts
   Avenue, charting attributes of thematic classes   151
   Changing elements   148
   Changing selected set   146
   Changing types   146
   Chart Properties dialog window   144
   Creation, from table subset   143
   Creation, method   143–168
   Data group   145
   Data marker, changing selected set   146
   Data marker, defined   144
   Data series   144
   Document categories   15
   Exercise   151–167
   Format limitations   146
   Frame, chart   173
   Identify tool, locating theme elements with   148
   Query Builder, changing selected set   146
   Ranges, charting attributes of thematic classes   150
   Select tool   148
   Typcs, changing   146
   Types, illustrated   147
Circles
   Adding   103
   Feature selection by shape   117
Classes
   Avenue, classes of objects   237
   Avenue, common inherited properties   237
   Avenue, objects as instances of class   237
   Avenue, script creation, Help system   242
   Class number, classification of theme   97
   Class values, classification of theme   97
Classification
   Class number   97

Class values 97
Equal interval method 96
Field choice 95
Legend, hiding 99
Legend, text editing 99
Method types 96
Quantile method 96
Ramp tool, color assignment by 97
Random tool, color assignment by 97
Summary statistics 98
Unique value method 96
Window 96
Clear Selected Features 284
Clear Selected Set icon 119
Client, Avenue communication with other applications, IAC 245
Close 284
Close All 285
Close Project 285
Color
   Changing 63
   Classes, color assignment 97
   Color Palette 95
   Hue 95
   Image data exercise 222
   Ramp, symbol color change exercise 107
   Saturation 95
   Symbol Editor, property changes with 32
   Tables, feature selection 115
   Value 95
Column Chart Gallery 285
Column chart, illustrated 147
Command strings, Avenue communication with other applications 244
Common map coordinate system 69–92
Compile button, Avenue script conversion 241
Completely Contain, theme-on-theme selection 190
Components, layout functions 170
Contain the Center Of, theme-on-theme selection 190

Continuous events 76
Contrast and brightness adjustment, image data exercise 222
Control Editor 234
Conversion
   ARC/INFO, CONVERTWORKSPACE command 58
   ArcView, converting spatial data to xiii
   Avenue script creation, converting with Compile button 241
   Exporting tables, converting to new format 205
   Logical queries on themes, converting to shape file 228
   Shape files, converting ARC/INFO coverages to 227
   Shape files, converting themes from ARC/INFO coverage 194
   Table field types, converting 202
Convert to Shapefile 194, 285
CONVERTWORKSPACE command 58
Copy 285
Copy Graphics 286
Copy Themes 286
Copying themes between views 81
Corporate use of ArcView, American Isuzu 254
Coverages, ARC/INFO 20
Create Chart 286
Create/Remove Index 286
Custom capture of spatial data, TIGER files 4
Custom output file creation, exporting tables 205
Customization with Avenue
   Avenue communication with other applications 244
   Avenue script creation, ArcView object class hierarchy 242
   Avenue script creation, converting with Compile button 241
   Avenue script creation, Customize Dialog window 242

Avenue script creation, debugging 241
Avenue script creation, Examine Variables button 242
Avenue script creation, Help system 242
Avenue script creation, linking with ArcView control 242
Avenue script creation, Script window 241
Avenue script creation, Search 242
Avenue script creation, steps 240–243
Avenue scripts, defined 238
Avenue scripts, syntax 240
Avenue start-up and shutdown scripts, customization 245
Avenue start-up and shutdown scripts, global changes 246
Avenue start-up and shutdown scripts, prompting the user 245
Avenue, arguments 238, 240
Avenue, as object oriented language 236
Avenue, class common inherited properties 237
Avenue, classes of objects 237
Avenue, object interaction by requests 237
Avenue, objects as instances of class 237
Avenue, properties of objects 237
Avenue, requests 240
Avenue, statements 238
Embedding permanent script changes 246
Encrypting permanent script changes 247
Graphical user interface customization, Customize dialog 234
Locking permanent script changes 247
Overview 7, xii
Permanent changes, embedding script 246
Permanent changes, encrypting script 247
Permanent changes, locking project 247
Customize 286
Customize dialog box 234
    Control Editor 234
    Delete button 234
    New button 234
    Properties List 234
    Pulldown lists 234
    Separator button 235
    Type list choices 234
Customize dialog window, Avenue script creation 242
Cut 287
Cut Graphics 287
Cut Themes 287

D

Data group, charts 145
Data marker, charts 144
Data queries
    Feature selection 115
    Feature selection by query 118
    Feature selection by shape 117
    Feature selection, locating selected set 120
    Fields Visible field controls 114
    Fields, display width 115
    Fields, names for 113
    Fields, resizing or reordering 114
    Logical queries on themes 120–123
    Sorting tables 115
    Table Properties window 113
    Views, tables with 115
Data series, charts 144
Database management skill, tables 119
DBase files
    Editing tables in ArcView 201
    Importing data to ArcView 25
    Tabular formats supported by Arcview 50
Debugging, Avenue script creation 241
Default project files, saving GUI customization 236
Default settings, Work Directory 20
Delete 287
Delete Button 234
Delete Field 287
Delete Graphics 288
Delete Records 288
Delete Themes 288
Desktop mapping application
    Automatic thematic maps 2
    Described 1

Origins 2–3
Destination table 72, 74
Detach Graphics 288
Digital elevation, Identity tool 222
Digital images 206
Directories, projects and views 53
Directory, sample project 20
Display, changing 187
Displaying data
    Classification, application exercise 107
    Classification, class number 97
    Classification, class number reduction exercise 109
    Classification, class values 97
    Classification, color assignment by Ramp tool 97
    Classification, color assignment by Random tool 97
    Classification, equal interval method 96
    Classification, field choice 95
    Classification, legend hiding 99
    Classification, legend text editing 99
    Classification, method types 96
    Classification, quantile method 96
    Classification, summary statistics 98
    Classification, unique value method 96
    Classification, window 96
    Exercise 106–107, 109, 111
    Feature identification, Identity tool 100
    Feature identification, Label tool 101
    Graphics, adding 102–103, 105
    Graphics, circles 103
    Graphics, Distance Units selector 103
    Graphics, Draw tool 103
    Graphics, editing 104
    Graphics, lines 103
    Graphics, Measure tool 103
    Graphics, moving 104
    Graphics, Pointer tool 104, 106
    Graphics, points 103
    Graphics, polygons 103
    Graphics, polygons reshaping 105
    Graphics, polyline reshaping 105
    Graphics, polylines 103
    Graphics, rectangles 103
    Graphics, reshaping 105
    Graphics, resizing 105
    Graphics, setting length or area 103
    Graphics, Symbol Palette 104–105
    Graphics, text additions 104
    Graphics, text label editing 106
    Graphics, text labels 102
    Graphics, Text tool 104
    Symbology, changing 93
    Symbology, Color Palette 95
    Symbology, defining 93
    Symbology, Fill Palette 94
    Symbology, Font Palette 94
    Symbology, Marker Palette 94
    Symbology, Pen Palette 94
    Text Labels Theme Properties window 102
Displaying data exercise
    Class number reduction 109
    Classification application 107
    Name of theme change 107
    Symbol color change 107
Distance Units selector 103
Distance, Measure tool 103
Distortion, map projections 132
Document categories, described 15
Documents, project organization, multiple documents 225
Draw 288
Draw tool 103, 117

E

Edit 288
Edit Legend 289
Editing shape files, shape files & hot links exercise 213
Embedding Avenue scripts 246
Encrypting Avenue scripts 247
Environmental Systems Research Institute. *See* ESRI
Equal interval method, classification of theme 97

Erase 289
Erase With Polygon 289
ESRI 103
    ARC/INFO's development 4
    GRID data 207
Event table 74
Event themes
    Address events 76
    Continuous events 76
    Creation 36
    Defined 74–79
    Described 37
    Event table 74
    Geocoding steps 78
    Linear events 76
    Naming conventions 80
    Point events 76
    Route events 76
    Shape file 74
    Tables, transforming into 85
    XY events 75
Examine Variables button, Avenue script creation 242
Exit 289
Export 289
Export Table 290
Exporting ARC/INFO data 57
Exporting tables
    Converting to new format 205
    Custom output file creation 205
    Saving joined tables 205
    Uses 205
Extending data
    Common map coordinate system 69–72
    Event themes 74–75, 77, 79
    Exercise 81, 83, 85, 87, 89, 91
    Joining tables 72–73
    Linking tables 74
    Research 91
Extending data exercise
    Event theme, transforming tables into 85
    Geocoded event theme, creation 88
    Geocoding edited tables 90

    Interactive address editing 90
    Links, establishing 81
    Matching tables 90
    Survey information, adding 87
    Tables, joining 82
    Themes, copying between views 81

F

Feature identification
    Identity tool 100
    Label tool 101
    Text labels 102
    Text Labels Theme Properties window 102
    Text labels, detaching 102
Features
    Hot link's function 193
    Shape files, merging features interactively 199
    Tables, feature selection 115
Fields
    Adding to table 157
    Classification of themes 95
    Editing tables, adding or deleting fields 202
    Editing tables, converting field types 202
    Editing tables, field values 201
    Editing tables, permanently deleting fields 202
    Index creation 226
    Resizing or reordering 114
    Shape files, adding fields 198
    Table creation from joined fields 125
    Tables, names for fields in 114
    Visible field controls 114
    Width 115
Files, locking to read-only 232
Fill Palette 94
Fill pattern, Symbol Editor 32
Find 290
Find tool 118
Fishbowl surveys, research 92
Five Ws of communication, cartographic design 177

Font Palette 94
Format
    Chart limitations 146
    Layout functions 169–188
Frame 290
Frame tool 172
Frames
    Changing frame properties 186
    Chart frame 173
    Creating with Frame tool 172
    Graphics frames, adding 182
    Layout functions 169
    Legend 172
    North arrow 173
    Picture frame 174
    Presentation/Draft display option 174
    Preserve View Frame box 172
    Scale bar 173
    Table frame 173
    Types 172
    View frames 172
    When Active/Always display option 174
Functionality reference 281–306

G

Geocoded event theme creation 88
Geocoding
    Edited tables 90
    Increasing matches 78
    Process described 40
    Steps 78
Geocoding Editor
    Dialog window, illustration 40
    Interactive address editing 90
Geographic coordinates, data formatting 69
Geographic information systems. *See* GIS
GIS
    ArcView's advantages xii
    Desktop mapping application 2
    Spatial mindset 8, 17–46
Government use of ArcView
    Arizona State Land Department 249
    Maine Department of Environmental Protection 251
Graphical user interface customization
    Avenue scripts, associating with controls 235
    Customize dialog box 234
    Customize dialog box, Separator button 235
    Customize dialog box, Type list choices 234
    Saving customization, default project file creation 236
    Saving customization, Save button for specific project 235
Graphics
    Adding 102–105
    Adding graphics to view 104
    Box tool 175
    Circles 103
    Distance Units selector 103
    Draw tool 103
    Editing 104
    Frames, picture 174
    Frames, supported graphics formats 174
    Graphic Size and Position window 176
    Graphic Size and Position window, moving with 176
    Graphics frames, adding 182
    Graphics Tool 197
    Group option 175
    Layouts, graphics page definition 170
    Length or area, setting 103
    Lines 103
    Measure tool 103
    Moving 104, 176
    Moving to background 175
    Pointer tool 104, 106
    Points 103
    Polygons 103
    Polygons reshaping 105
    Polyline reshaping 105
    Polylines 103
    Rectangles 103

Index

Reshaping 105
Resizing 105
Shape files 194
Shape files, merging graphics 199
Symbol Palette 104–105
Table feature selection by shape 117
Text additions 104
Text label editing 106
Text labels 102
Text tool 104, 175
GRID data
    Controlling display 207
    Defined 207
    Identity tool 222
    Scale, setting 207
Grids
    Hiding the grid 175
    Layout Properties window 175
    Map grid 175
    Snapping grid 175
    Spacing of grid 175
Group 290
Group option 175

H

Have their Center In, theme-on-theme selection 190
Help system
    Avenue script creation 242
    HelpTool icon illustation 47
    Menu illustration 47
Hidden fields in tables 114
Hide/Show Legend 290
Hiding legend 99
Hot Link 291
Hot links
    Avenue script execution 193
    Defining 193
    Exercise 208–217
    Function 193
    Hot Link tool 193
    Image formats supported 193
    Predefined action 193

Project organization, multiple views 225
Project organization, view-of-views 228
Hue 95

I

Icons
    Add Event Themes icon 36
    Add Theme icon, illustration 20
    Clear Selected Features 119
    Clear Selected Set 119
    Color Palette 95
    Described 14
    Fill Palette 94
    Font Palette 94
    HelpTool icon 47
    Marker Palette 94
    Open Attribute Table for Active Theme icon, illustrated 28
    Pen Palette 94
    Promote 117
    Save Project 56
    Select All 119
    Series From Records/Series From Fields 155
    Sort Ascending 115
    Sort Descending 115
    Switch Selected Set 119
    Zoom In icon, illustration 23
    Zoom Out icon, illustration 23
    Zoom to Active Themes 24
    Zoom to Full Extent 24
    Zoom to Selected Features 120
Identify 291
Identify tool, locating theme elements in charts 148
Identity tool
    Image data exercise 222
    Theme attribute examination 100
Image data
    Aerial imagery 206
    Controlling display 207
    Digital images 206
    Exercise 218–221

Manipulation tools 207
Scale, setting 207
Types 206
Image data exercise
  Color 222
  Contrast and brightness adjustment 222
  Identity tool 222
  Interval tool 222
  Opening project and view window 218–219, 221
Image formats, hot links 193
Import 291
IMPORT utility 58
Importing data
  ARC/INFO data 57
  Site selection sample project 25
  Tabular data, adding to project 50
INFO files
  Editing tables in ArcView 201
  Importing data to ArcView 25
INFO tables, tabular formats supported by Arcview 50
Informix, tabular formats supported by Arcview 50
Installation and configuration
  Memory requirements 277
  Operating systems 279
  Pre-release versions of ArcView and 32-bit extensions 278
Inter-application communication in Avenue 244
Interactive address editing 90
Intersect
  Spatial join 192
  Theme-on-theme selection 190
Interval tool, image data exercise 222

J

Join 291
Joining data files to spatial data themes 27
Joining tables
  Chart exercise, joining table to theme 152
  Destination table 72, 74
  Editing joined tables 201
  Exercise 82
  Exporting joined tables 205
  Illustration 29
  Join function 72
  Joined tables, illustration 30
  Order 82
  Reusable applications, tabular data organization 224
  Source table 72, 74
  Start-up time reduction 229
  Virtual table 72
Joining themes 125

L

Label 291
Label tool 101
Labels, text 102
Landscape format, Page Setup dialog box 180
Latitude & longitude
  Common map coordinate system 69
  Geocoding 75
Layout 292
Layout exercise
  Box tool 187
  Changing frame properties 186
  Creating new layout 184
  Display, changing 187
  Graphics frames, adding 182
  Neatline, adding 187
  Opening stored template 184
  Printing 188
  Template, storing as 182
  Text tool 187
  Text, adding 187
Layout Properties window 175
Layouts
  Creation 170–171
  Defined 169—188
  Document categories 15
  Exercise 180–181, 183, 185, 187
  Frames, chart frame 173

Frames, creating with Frame tool   172
Frames, north arrow   173
Frames, picture frame   174
Frames, Presentation/Draft display option   174
Frames, Preserve View Frame box   172
Frames, scale bar   173
Frames, scale parameters of view frame   172
Frames, table frame   173
Frames, types   172
Frames, view frames   172
Frames, When Active/Always display option   174
Functions   169—188
Graphics page definition   170
Map composition, cartographic design   176
Map composition, graphics   175–179
Map composition, positioning map elements   175
Map elements addition   170
Printing   180
Printing, Page Setup dialog box   180
Legend Editor
    Access   63
    Display options   93
    Window, illustration   96
Legends
    Editing   99
    Frames   173
    Hiding   99
Length of graphics   103
Libraries   53
Line Chart Gallery   292
Line chart, illustrated   147
Line in line spatial joins   192
Line in polygon spatial joins   192
Line to point spatial joins   192
Linear events   76
Lines, adding   103
Link   292
Linking tables, Link function   74

Links
    Avenue script creation, linking with ArcView control   242
    Extending data exercise   81
    See Hot links. See Hot links
    View frames   172
Live links   172–173
Locate   292
Locking projects, Avenue script   247
Locking theme and project   231
Logical queries
    Table feature selection by query   118–119
    Tabular data, converting to shape file   229
Logical queries on themes
    Defining   121
    Described   121
    Query Builder   121
    Theme Query Builder dialog window   121

M

Main ArcView window   13
Maine Department of Environmental Protection's use of ArcView   251
Maintenance
    ARC/INFO coverages   230
    Importing and exporting data   229
    Shape files   230
Make Default button, saving GUI customization   236
Map composition
    Cartographic design, 5 Ws of communication   177
    Cartographic design, defined   176
    Cartographic design, style development   178
    Graphics, Box tool   175
    Graphics, Group option   175
    Graphics, moving to background   175
    Graphics, Text tool   175
    Positioning map elements, Align tool   175

Positioning map elements, Graphic Size and Position 176
Positioning map elements, hiding the grid 175
Positioning map elements, Layout Properties window 175
Positioning map elements, moving graphics 176
Positioning map elements, snapping grid 175
Map coordinates, extending data exercise 84
Map grid 175
Map projections
    Data formatting 69
    Table use exercise 130
Map scale 66
Margins
    Page Setup dialog box 180
    Positioning map elements, Align tool 175
Marker Palette 94
Matching tables 90
Measure 292
Measure tool 103
Memory requirements, installation and configuration 277
Merge Graphics 293
Merge operations, shape files & hot links exercise 211
MicroVision segments 136, 152, 307–308
Mouse
    Moving graphics 176
    Table feature selection, methods 115
Multiple documents, project organization 225
Multiple views, project organization 225

N

Names
    Alias for fields in tables 113
    Displaying data exercise 107
    Syntax, Avenue scripts 240
Naming conventions, event themes 80
Nearest, spatial join 192
Neatline, adding 187
New button 234
New Project 293
New Theme 293
North arrow frames 173

O

Objects. See Avenue 237
Open Attribute Table for Active Theme icon, illustrated 28
Open Project 293
Open Theme Table 293
Opening project
    Data sources 50
    Exercise 57–67
    Opening view 49
    Raster data, adding 49
    Spatial data, adding 49
    Steps 48–49, 51
    Tabular data, adding 50
    Vector data, adding 49
Opening project exercise
    ARC/INFO data, importing and exporting 57
    Spatial and tabular data, loading 57, 60
Opening view
    Sample project 20
Operating systems, installation and configuration 279
Oracle, tabular formats supported by Arcview 50
Ordering
    Fields, reordering 114
    Tables, reordering 117
Output resolution, Page Setup dialog box 180
Overlay operations, Theme-on-theme selection 189

P

Page format, layout functions 169
Page Setup 293
Page Setup dialog box 180

Palette Editor access 63
Pan 294
Pan tool, described 24
Paste 294
Pen Palette 94
Pie Chart Gallery 294
Pie chart, illustrated 149
Pin studies, research 91
Point events 76
Point in polygon spatial joins 192
Point to line spatial joins 192
Point to point spatial joins 192
Pointer 294
Pointer tool 104, 106
Points, adding 103
Polygon in polygon spatial joins 192
Polygons
    Adding 103
    Feature selection by shape 117
    Reshaping 105
Polylines
    Adding 103
    Feature selection by shape 117
    Reshaping 105
Portrait format, Page Setup dialog box 180
Positioning map elements
    Align tool 175
    Graphic Size and Position window 176
    Hiding the grid 175
    Layout Properties window 175
    Map grid 175
    Moving graphics 176
    Snapping grid 175
Pre-release versions of ArcView and 32-bit extensions 278
Presentation/Draft display option 174
Preserve View Frame box 172
Print 294
Print Setup 295
Printing
    Layout exercise 188
    Layouts 180
    Page Setup dialog box 180
    To file 180
Project organization
    Multiple documents 225
    Multiple views 225
Project pull-down menu 53
Project window, described 14
Projection properties, shape files & hot links exercise 209
Projects and views
    Add Table dialog window 52
    Add Tables dialog window 53
    Add Theme 52
    Add Theme dialog window 52, 54
    Adding data 52–53
    Data sources 50
    Directories 53
    Help menu 48
    Libraries 53
    Opening project, exercise 57–67
    Opening project, steps 48–49, 51
    Project pull-down menu 53
    Project window, accessing data from 53
    Saving work 56
    Themes, color change 63
    Themes, creation 55
    Themes, defined 54
    Themes, display 62
    Themes, drawing order change 62
    Themes, Legend Editor access 63
    Themes, making active 56
    Themes, organized by views 55
    Themes, Palette Editor access 63
    Themes, setting properties 55
    Themes, stored using latitude and longitude 66
    Views, adding themes 66
    Views, display 66
    Views, map scale 66
    Views, organizing themes 55
    Views, setting properties 56
    Views, theme drawing order 55
    Views, window contents 55
Promote 295

Promote tool 117
Properties
    Avenue, class common inherited
        properties 237
    Avenue, properties of objects 237
    Properties List 234
    Properties of objects in Avenue 237
    Themes, setting properties 55
    Views, setting properties 56
Properties–Graphic 295
Properties–Layout 295
Properties–Project 295
Properties–Table 296
Properties–View 296
Pull-down lists 234
Pull-down menus, described 14

Q

Quantile method, classification of theme 97
Queries
    Indexes, improving performance of fields 226
Query 296
Query Builder 118
    Charts, changing selected set 146
    Logical queries on themes 121
    Theme Query Builder dialog window 121

R

Ramp tool, class color assignment by 97
Random tool, class color assignment by 97
Raster data, adding 49
Raster-based data format, GRID data 207
Rectangles, adding 103
Refresh 296
Remove All Joins 297
Remove All Links 297
Rename 297
Reordering
    Fields 114
    Tables 117
Repositioning, text labels 106

Requests
    Avenue 240
    Avenue script creation, Help system 242
    Avenue, object interaction by 238
Reset button, saving GUI customization 236
Reshaping graphics 105
Resizing
    Fields 114
    Graphics 105
    Text labels 106
ReMatch 296
Reusable applications
    Locking theme and project 231
    Maintenance ease, ARC/INFO coverages 230
    Maintenance ease, importing and exporting data 229
    Maintenance ease, shape files 230
    Maintenance ease, tabular data 231
    Project organization, index creation 226
    Project organization, multiple documents 225
    Project organization, multiple views 225
    Project organization, performance optimization 226
    Shape files, converting ARC/INFO coverages to 227
    Spatial data organization 223
    Start-up time reduction, need for 228
    Start-up time reduction, reducing logical queries on 228
    Tabular data organization 224
Route events 76

S

Sample project
    ArcView overview 18
    Site selection. See Site selection sample project 18
Saturation 95
Save Project 297

Save Project As   297
Saving work
   Customization, Default project file creation   236
   Customization, Save button for specific project   235
   Projects and views   56
   Save As, locking projects   232
   Save Project As   56
   Save Project icon   56
Scale
   Frames, legend   172
   Frames, setting parameters of view frame   172
   Maps   66
Scale bar frames   173
Script window   241
Search, Avenue script creation   242
Select   298
Select All   298
Select All Graphics   298
Select All icon   119
Select By Theme   190, 298
Select Feature   298
Select Features Using Shape   298
Select None   299
Select tool   148
Selected set, changing charts   146
Send to Back   299
Separator button   235
Series From Records/Fields   299
Series From Records/Series From Fields icon   155
Server, Avenue communication with other applications, IAC   245
Shape files   74
   ARC/INFO, converting coverages to   227
   Attributes for geographic features   194
   Convert to Shape file   194
   Converting themes from ARC/INFO coverage   194
   Editing, Graphics Tool   197
   Editing, permanent changes   197
   Editing, steps for   195
   Exercise   208–217
   Graphics   194
   Logical queries on themes, converting   228
   Maintenance of project   230
   Merging features by attribute value   200
   Merging features by spatial aggregation   200
   Merging features interactively   199
   New themes, adding attributes   198
   New themes, adding fields   198
   New themes, creating   198
   Spatial data storage function   194
   Theme snapping properties   197
Shape files & hot links exercise
   Changing features to shape file   210
   Creating shape file   210
   Editing shape files   213
   Hot link creation   216
   Merge operations   211
   Projection properties   209
Shape, table feature selection   117
Show Symbol Palette   300
Show/Hide Grid   299
Show/Hide Legend   299
Show/Hide Margins   300
Show/Hide Title   300
Show/Hide X Axis   300
Show/Hide Y Axis   300
Site selection sample project   18
   Add Event Themes dialog box, illustration   38
   Add Event Themes icon   36
   Add Theme icon   20
   Adding tables   26
   Adding themes to view   21
   Address matched event theme   41
   Applied classification on theme, illustration   32
   Attributes of Tracts table, illustrated   28
   Census tract theme, illustration   23
   Directory, setting   20
   Event theme described   37

Event theme, creation 36
Geocoded event theme, saving 39–40
Geocoding Editor dialog window,
   illustration 40
Importing tabular data sets 25
Joined tables, illustration 30
Joining, data files to spatial data themes 27
Joining, tables prepared for 29
Matchable theme, steps for making 33
Open Attribute Table for Active Theme
   icon, illustrated 28
Opening 20
Pan tool 24
Second field, adding 43
Symbol Editor, property changes with 32
Thematic map, creation 30
Themes, adding 20
TIGER, correcting errors and omissions in
   35
View with themes added 22
Zoom In 23
Zoom Out 23
Zoom to Active Themes 24
Zoom to Full Extent 24
Size and Position 301
Sizing
   Fields, resizing 114
   Graphics, resizing 105
   Page size, Page Setup dialog box 180
   Text labels, resizing 106
Snap 301
Snapping grid 175
Sort Ascending 115
Sort Ascending / Sort Descending 301
Sort Descending 115
Sorting tables 115
Source table 72, 74
Spatial aggregation, shape files, merging
   features 200
Spatial data
   Adding 49
   Loading 57
   Organization, reusable applications 224

Spatial joins 124
   Function 192
   Queries 192
   Relation types 192
   Steps to perform 192
   Types 192
Spatial queries 192
SQL
   Tables and database management skill 119
   Tabular formats supported by ArcView 50
SQL Connect 301
Standard Industrial Classification codes
   141
Start-up time reduction
   Logical queries on themes, converting to
      shape file 228
   Reusable applications 228
Start/Stop Editing 301
State Plane Coordinate, common map
   coordinate system 70
Statements in Avenue 238
Static links 172–173
Statistics 302
   Classification of theme 98
   Tables, generating statistics for 203
   Tables, summary statistics 204
Status bar, described 14
Store As Template 302
Summarize 302
Summarize tool 204
Summary statistics, classification of theme
   98
Summary table creation, chart exercise 156
Survey information, extending data
   exercise 87
Switch Selected Set icon 119
Switch Selection 302
Sybase, tabular formats supported by
   ArcView 50
Symbol Editor, property changes with 32
Symbol Palette 105
Symbology
   Changing 93

Color Palette 95
Defining 93
Fill Palette 94
Font Palette 94
Legend Editor 93
Marker Palette 94
Pen Palette 94
Syntax, Avenue scripts 240
System.Execute, Avenue communication with other applications 244

T

Table of contents
Hiding legend 99
Legend frames 172
Table use exercise
Auto-Label tool 138
Joining tables 129
Joining themes 124
Map projections 130
MicroVision segmentation data 136
Spatial join 124
Standard Industrial Classification codes 141
Table creation from joined fields 125
Tables
ArcView's template 113
Attribute table creation 161
Attributes of Tracts table, illustrated 28
Chart creation from table subset 143
Classification of fields 95
Database management skill 119
Destination table 72, 74
Document categories 15
Editing, adding or deleting fields 202
Editing, field values 201
Editing, joined tables 201
Editing, permanently deleting fields 202
Editing, types editable in ArcView 201
Event table 74
Exercise 124–141

Exporting tables, converting to new format 205
Exporting tables, saving joined tables 205
Exporting tables, uses 205
Feature selection by query 118–119
Feature selection by query, Clear Selected Features 119
Feature selection by query, Find tool 118
Feature selection by query, Query Builder 118
Feature selection by query, Switch Selection 119
Feature selection by shape, Draw tool 117
Feature selection by shape, graphics creation 117
Feature selection, color 115
Feature selection, locating selected set 120
Feature selection, logical queries on themes 121
Feature selection, methods 115
Feature selection, Promote tool 117
Feature selection, reordering table 117
Feature selection, Zoom to Selected tool 120
Field types, converting 202
Fields, display width 115
Fields, resizing or reordering 114
Frames, table 173
Geocoding edited tables 90
Index creation 226
Joined tables, illustration 30
Joining 82
Joining tables 72–73
Joining, tables prepared for 29
Link function 74
Logical queries on themes, defining 121
Logical queries on themes, Query Builder 121
Logical queries on themes, Theme Query Builder dialog 121
Maintenance of project 231
Names for fields 113
Site selection sample project 26

Sorting 115
Source table 72, 74
Statistics, generating 203
Statistics, joining summary to primary table 205
Statistics, summary statistics 204
Summary table creation 163
Table Properties window 113
Views, tables with 115
Virtual table 72
Visible field controls 114
Templates
   Layout functions 170
   Layout, storing as template 182
   Tables 113
Text 302
   Adding 187
   Graphics, adding to 104
Text labels
   Editing 106
   Repositioning 106
   Resizing 106
Text Labels Theme Properties window 102
Text tool 104, 175, 187
Thematic map, creation 30
Theme attribute tables
   Joining tables, order 84
   Reusable applications, tabular data organization 225
Theme elements, locating on charts 148
Theme Properties 303
Theme snapping properties, shape files 197
Theme-on-theme selection
   Methods of selection 191
   Select By Theme 190
   Spatial relation types 189
Themes
   Add Theme dialog window 54
   Add Theme icon 21
   Adding themes to view 21
   Address matched event theme 41
   Census tract theme, illustration 23

Chart exercise, joining table to theme 152
Charting attributes of thematic classes 150
Color change 63
Copying between views 81
Creation 55
Defined 20
Display 62
Drawing order change 62
Drawing order in views 55
Event theme described 37
Event theme, creation 36
Event themes 74–79
Geocoded event theme 41
Legend Editor access 63
Locking theme 231
Making theme active 56
Matchable theme, steps for making 33
Palette Editor access 63
Setting properties 55
Shape files, creating new themes 198
Spatial join 124
Stored using latitude and longitude 66
Tables, logical queries on themes 121
Text labels 102
Text Labels Theme Properties window 102
Theme Properties window, locking theme 231
Themes On / Themes Off 303
TIGER 19
TIGER files 4
   Correcting errors and omissions in 35
   Geocoding, increasing matches 80
Tolerances, snapping 197
Topographically Integrated Geographically Encoded Reference. *See* Files; TIGER files
Type list choices 234

U

Undo Erase 303
Ungroup 303
Unique value method, classification of theme 97

Universal Transverse Mercator, common map coordinate system 70
University of Missouri at St. Louis' use of ArcView 255
University of Wisconsin-Madison's use of ArcView 257
UNIX xii
   ArcView 8
   Opening project 48–51
Use Template 303

V

Value 95
Vector data, adding 49
View frames 172
Views
   Adding themes 66
   Display 66
   Document categories 15
   Hot link's function 193
   Locking theme 231
   Map scale 66
   Organizing themes 55
   Project organization, multiple views 225
   Setting properties 56
   Table of contents, legend frames 172
   Tables with 115
   Theme drawing order 55
   View-of-views, project organization 228
   Window contents 55
Virtual table 72

Visible field controls 114

W

When Active/Always display option 174
Width
   Fields 115
   Symbol Editor, property changes with 32
Work Directory, default settings 20

X

XY events 75
XY Scatter Chart Gallery 304
XY scatter chart, illustrated 147

Z

Zoom In 304
   Avenue scripts 238
   Icon, button, and Tool 24
Zoom In (Tool) 304
Zoom Out 304
Zoom Out (Tool) 304
Zoom Out icon, button, and Tool 24
Zoom to Active Themes icon 24
Zoom to Actual Size 305
Zoom to Full Extent 305
Zoom to Full Extent icon 24
Zoom to Page 305
Zoom to Selected 305
Zoom to Selected Features 120
Zoom to Themes 305

More OnWord Press Titles

Pro/ENGINEER and Pro/JR. Books

INSIDE Pro/ENGINEER
Book $49.95 Includes Disk

Pro/ENGINEER Quick Reference, 2d ed.
Book $24.95

Pro/ENGINEER Exercise Book
Book $39.95 Includes Disk

Thinking Pro/ENGINEER
Book $49.95

INSIDE Pro/JR.
Book $49.95

Interleaf Books

INSIDE Interleaf
Book $49.95 Includes Disk

Adventurer's Guide to Interleaf Lisp
Book $49.95 Includes Disk

The Interleaf Exercise Book
Book $39.95 Includes Disk

The Interleaf Quick Reference
Book $24.95

Interleaf Tips and Tricks
Book $49.95 Includes Disk

MicroStation Books

INSIDE MicroStation 5X, 3d ed.
Book $34.95 Includes Disk

MicroStation Reference Guide 5.X
Book $18.95

MicroStation Exercise Book 5.X
Book $34.95
Optional Instructor's Guide $14.95

MicroStation 5.X Delta Book
Book $19.95

MicroStation for AutoCAD Users, 2d ed.
Book $34.95

Adventures in MicroStation 3D
Book 49.95 Includes Disk

MicroStation Productivity Book
Book $39.95
Optional Disk $49.95

MicroStation Bible
Book $49.95
Optional Disks $49.95

Build Cell
Software $69.95

101 MDL Commands
Book $49.95
Optional Executable Disk $101.00
Optional Source Disks (6) $259.95

101 User Commands
Book $49.95
Optional Disk $101.00

Bill Steinbock's Pocket MDL Programmer's Guide
Book $24.95

Managing and Networking MicroStation
Book $29.95
Optional Disk $29.95

The MicroStation Database Book
Book $29.95
Optional Disk $29.95

INSIDE I/RAS B
Book $24.95 Includes Disk

The CLIX Workstation User's Guide
Book $34.95 Includes Disk

SunSoft Solaris Series

The SunSoft Solaris 2.* User's Guide
Book $29.95 Includes Disk

SunSoft Solaris 2.* for Managers and Administrators
Book $34.95

The SunSoft Solaris 2.* Quick Reference
Book $18.95

Five Steps to SunSoft Solaris 2.*
Book $24.95 Includes Disk

One Minute SunSoft Solaris Manager
Book $14.95

SunSoft Solaris 2.* for Windows Users
Book $24.95

The Hewlett Packard HP-UX Series

The HP-UX User's Guide
Book $29.95 Includes Disk

The HP-UX Quick Reference
Book $18.95

Five Steps to HP-UX
Book $24.95 Includes Disk

One Minute HP-UX Manager
Book $14.95

CAD Management

One Minute CAD Manager
Book $14.95

Manager's Guide to Computer-Aided Engineering
Book $49.95

Other CAD

CAD and the Practice of Architecture: ASG Solutions
Book $39.95 Includes Disk

INSIDE CADVANCE
Book $34.95 Includes Disk

Using Drafix Windows CAD
Book $34.95 Includes Disk

Fallingwater in 3D Studio: A Case Study and Tutorial
Book $39.95 Includes Disk

Geographic Information Systems/ESRI

The GIS Book, 3d ed.
Book $34.95

INSIDE ARC/INFO
Book $74.95 Includes CD

ARC/INFO Quick Reference
Book $24.95

ArcView Developer's Guide
Book $49.95

INSIDE ArcView
Book $39.95 Includes CD

DTP/CAD Clip Art

1001 DTP/CAD Symbols Clip Art Library: Architectural
Book $29.95

DISK FORMATS:
MicroStation
  DGN Disk $175.00
  Book/Disk $195.00

AutoCAD
  DWG Disk $175.00
  Book/Disk $195.00

CAD/DTP
  DXF Disk $195.00
  Book/Disk $225.00

OnWord Press Distribution

End Users/User Groups/Corporate Sales

OnWord Press books are available worldwide to end users, user groups, and corporate accounts from your local bookseller or computer/software dealer, or from HMP Direct: call 1-800-223-6397 or 505-473-5454; fax 505-471-4424; write to High Mountain Press Direct/Softstore, 2530 Camino Entrada, Santa Fe, NM 87505-8435, or e-mail to ORDERS @BOOKSTORE.HMP.COM.

Wholesale, Including Overseas Distribution

High Mountain Press distributes OnWord Press books internationally. For terms call 1-800-4-ONWORD or 505-473-5454; fax to 505-471-4424; e-mail to ORDERS@ IPG.HMP.COM; or write to High Mountain Press/IPG, 2530 Camino Entrada, Santa Fe, NM 87505-8435, USA. Outside North America, call 505-471-4243.

Comments and Corrections

Your comments can help us make better products. If you find an error in our products, or have any other comments, positive or negative, we'd like to know! Please write to us at the address below or contact our e-mail address: READERS@HMP.COM.